AF876863

Gianfranco Dioguardi
Politecnico Università di Bari
Piazza Eroi del Mare, 9
70121 Bari
Italy
dioguardi@poliba.it

ISBN 978-1-4419-1331-9 e-ISBN 978-1-4419-1333-3
DOI 10.1007/978-1-4419-1333-3

Library of Congress Control Number: 2009937879

© Gianfranco Dioguardi, 2010
Originally published in Italian as *Le imprese rete*, © Bollati Boringhieri editore, Torino, 2007
All rights reserved. This work may not be translated or copied in whole or in part without the written permission of the publisher (Springer Science+Business Media, LLC, 233 Spring Street, New York, NY 10013, USA), except for brief excerpts in connection with reviews or scholarly analysis. Use in connection with any form of information storage and retrieval, electronic adaptation, computer software, or by similar or dissimilar methodology now known or hereafter developed is forbidden.
The use in this publication of trade names, trademarks, service marks, and similar terms, even if they are not identified as such, is not to be taken as an expression of opinion as to whether or not they are subject to proprietary rights.

Printed on acid-free paper

Gianfranco Dioguardi

Network Enterprises

The Evolution of Organizational Models from Guilds to Assembly Lines to Innovation Clusters

English Translation by Rachel Barrit Costa

Springer

Innovation, Technology, and Knowledge Management

Series Editor
Elias G. Carayannis, George Washington University, Washington D.C., USA

For further volumes:
http://www.springer.com/series/8124

Dedication

Dedication, that is, a premise whose role is to be a sign of esteem or affection for a particularly important or authoritative person or a person who is dear to us. Ancient texts very often feature a dedication, above all texts on scientific matters, and I have been filled with curiosity at these introductory tributes ever since my first acquaintance with bibliophily. This was on the occasion of my encounter with an important 18th century Neapolitan edition – published in 1714 by Bernardo-Michele Raillard – of the celebrated essays on natural science experiments conducted at the Academy of Cimento, entitled *Saggi di naturali esperienze fatte nell'Accademia del Cimento* (1667) [Natural Experiments Conducted at the Academy of Cimento] by Lorenzo Magalotti (1637–1712), who was the secretary of that eminent scientific association. I bought the book because I was fascinated by its sumptuous appearance in folio and by the rapturous praise it received from Elisabeth Seacombe, of the Roman antiquarian bookshop "*Rappaport*." But I was equally fascinated by the motto of the Cimento Academy "TRY AND TRY AGAIN," which was cited in numerous places by the author.

Precisely that motto was one of my favorites as a useful guide to the "business of living"; furthermore, the text on experimental science, which was most absorbing, formed an excellent accompaniment to the valuable bibliophilic edition. The work thus acted as my mentor, thanks also to the words of wisdom of Mrs. Seacombe, who led me into the wondrous magic world of antique books, the love of which I would cherish forever after.

It was the dedication placed as a foreword to the book that particularly attracted me, with its indication that the academy lived "*Under the Protection of the Most Serene and Noble Prince Leopold of Tuscany.*" The work was enhanced by the explicit dedication to "*His Serene Excellency Don Cesare Michel-Angelo Davalos D'Aquino, D'Aragona,*" followed by a string of noble titles. There was also a "*Proem to the Readers,*" in other words a sort of preface or preliminary explanation for those who wished to delve into the collection of essays.

From the very start my curiosity as a bibliophile constantly prompted me to browse through books, especially antique books, in search of this kind of preface, even if it might not be formally declared as an integral part of the text itself.

Such is the case, for instance, of *De Rerum Natura* by Titus Lucretius Carus, the Latin poet of whose life only sketchy details are known. Veiled in legend, information on his life story was handed down by St. Jerome, according to whom the poet was born in 96 or 94 BC, committing suicide barely 44 years later, i.e., in 53 or 51 BC. *De Rerum Natura*, in the version that has come down to us, revised apparently by Cicero, does not include an explicit dedication, but in the major portion of Book I – not forming part of the text, according to some – Lucretius weaves a hymn to Venus, then proceeds to present the hero of this poem, namely the Greek philosopher Epicurus.

Lucretius opens his poem in terms which – it seems to me – are strikingly relevant to the contemporary age: "Mother of Rome, delight of Gods and men/Dear Venus [...]/Lull to a timely rest/O'er sea and land the savage works of war" (I, 2, 29, 30). Then there is a genuine dedication to Epicurus: "Whilst human kind/Throughout the lands lay miserably crushed/Before all eyes beneath Religion- who/Would show her head along the region skies/Glowering on mortals with her hideous face-/A Greek it was who first opposing dared/Raise mortal eyes that terror to withstand/Whom nor the fame of Gods nor lightning's stroke/Nor threatening thunder of the ominous sky/Abashed; but rather chafed to angry zest/His dauntless heart to be the first to rend/The crossbars at the gates of Nature old." (I, 62–71)

The lines cited appear in the translation of *De Rerum Natura* by William Ellery Leonard.[1]

The explicit use of dedications became widespread with the advent of printing, but their period of greatest triumph was in the Renaissance and the following centuries. It was a trend that was immediately adopted by the great scientists such as Galileo Galilei (1564–1642). For example, his *Sidereal Messenger* (1610) opens with a long dedication to "*His Serene Highness Cosimo II of the Medici IV Grand-Duke of Tuscany*," to whom, when the prince was still the young Prince Royal, Galileo himself had given some lessons in mathematics, upon the request of Cosimo's mother Cristina of Lorraine, thereby gaining her profound affection. Moreover, Galileo's most important and celebrated *Dialogue Concerning the Two Chief World Systems* (1632), defined in the introduction to the work by Francesco Flora as "the greatest masterpiece of scientific literature of all time and of all nations,"[2] was likewise dedicated to the Medici family, namely to Ferdinand II of the Medici Grand-duke of Tuscany, who had succeeded Cosimo II in 1621. The dedication was followed by a section entitled "To the Discreet Reader," which acts as a proem to the entire work.

The custom of including a dedication continued into the enlightened 18th century, where Denis Diderot (1713–1784), one of the most significant interpreters of the age of Enlightenment, would offer a similar composition – even though in impersonal terms – in one of his most important treatises. I refer here to the *Thoughts on the Interpretation of Nature*, a small work devoted to the scientific method that drew its inspiration from Bacon's *Novum Organum [The New Organon]*. It was published in a first and extremely rare edition in 1753, and subsequently in two further editions, which were enlarged in 1754. Diderot's essay is dedicated to "*Young People Who Are Setting out on the Study of Natural Philosophy*" and he clarifies the aims of his work without further ado, addressing young scholars with the statement, among other things, that "Given that, rather than instructing you it is my aim to exercise your mind, I am little concerned with the question of whether you accept or reject my ideas, provided that they engage your full attention. A more knowledgeable person will teach you to grasp the forces of nature; I will rest content with having made you put your own to the test."[3] These are concepts I wholeheartedly endorse in presenting this book on *The Network Enterprise*, and I express my equally wholehearted endorsement of other statements by the French philosopher, likewise taken from *The Interpretation of Nature*, for example, where he writes "7. As long as things remain merely within our understanding, they are our opinions; they are notions that can be true or false, granted or contradicted. They acquire consistency only by becoming linked to external beings. This link is achieved either through an uninterrupted chain of experiences, or through an uninterrupted chain of arguments that depend on the one hand on observation, and on the other on experience, or else through a chain of experiences scattered here and there amidst the arguments."[4]

I have endeavored to follow the Diderotian teaching, which has led me to "try and try again" to observe the experiences connected to the "external beings" – in this specific case, entrepreneurs – in order to try to generalize the observations by constructing a theory. This accounts for the gratitude I feel toward those to whom I owe the prime cause of inspiration for this book.

I will therefore take up again the ancient custom of the dedication.

This book on *The Network Enterprise* is dedicated to an important entrepreneur who is dear to me, exactly as suggested by the aforementioned semantic explanation of dedications. It is dedicated to Claudio Navarra, who set up a prominent entrepreneurial group in the economic sector of the Built Environment, in which the *Italiana Costruzioni Company* is the leading firm and whose brilliant CEOs today are his sons Attilio and Luca, with the meticulous consultancy of Benedetta Navarra.

I met Claudio Navarra during the period when I too was working in the same sector of constructions. He proved to be a challenging but extremely principled competitor. I was thus able to engage in a relationship of profound and stimulating friendship. Later I was lucky enough to have the opportunity to work directly with him and with his group, which gave me solid experience in the field. I was thereby also able to perfect the experimental phase of how to be an entrepreneur and how to work in the world of business according to the dictates of intuition and the principles of organization which come to the fore, essentially, in the so-called enterprise spirit. This is why it is my desire today to dedicate this book to Claudio Navarra, who indirectly but fundamentally contributed to its composition.

The book grew out of a revisitation of the useful experiences described above: reflecting on those earlier days, I was able to draw a more general message that developed into a theoretical construct in its own right. This evolution was further enhanced by my teaching activity at the Faculty of Engineering at the Bari Politecnico University, in my course on *Economics and Business Organization*. As is typically the case with course design, preparation of a series of lectures requires a preliminary analysis of the various different experiences, so that they can be transmitted, during the lessons, with the connotation and dignity of a sound theory.

The forging of a theoretical framework is further aided by the constant dialectical interaction that takes place on a daily basis through in-depth communication with students and colleagues. It was in this manner that *The Network Enterprise* gradually took shape.

In accordance with the natural line of development of a dedication, I would now like to turn to the important moment of acknowledgments and thanks: directly to the students themselves who attended my course and, naturally, to my colleagues and coworkers with whom I engage in exchange of ideas day by day.

I would also like to mention and thank all the students, metaphorically one by one, who since 1971 have succeeded each other year after year as enrollees in my course. I would also like to express my thanks to the colleagues who have worked and are working with me at the Department of Engineering Management and who jointly have established the "Bari School" of Business Organization: Vito Albino, Nicola Costantino, Michele Gorgoglione and likewise Claudio Garavelli, Pier Paolo Pontrandolfo, and the delightful colleagues Nunzia Carbonara, Ilaria Giannoccaro, Barbara Scozzi, and all the others. Many of them have now become tenured professors and significant protagonists in the Italian world of science and worldwide science. Others are following on the challenging path that will lead them to the highest ranks of their university career.

To all of them I would like to convey my sincerest feelings of profound gratitude for the inspiration that supported me in composing this book, which without their contribution and their dialectical contribution probably would never have been created.

Let me also take this opportunity to mention Maurizio Alfieri, Marilde Longo, Enzo Maiorano, and Ferdinando Napoli: these graduates in engineering, no sooner had they completed their degree at the Politecnico University of Bari, entered the corporate world by launching the *Edilportale* company, an enterprise that has achieved considerable success in the field of the New Economy. They represent a concrete testimony of the potential of university teaching to stimulate, in those who have aptitude and flair, the resolve to set up successful enterprises.

As an afterword to the "Dedication," together with the expression of thanks, I will offer a brief "Proem," as was customary in the ancient texts I have had the pleasure and the good fortune to peruse.

The book is shaped as a "lesson," inasmuch as it has arisen precisely from the lessons I held in Bari during the course in *Economics and Business Organization.*

A historical analysis of the world of organizations and the theories that have forged their development forms the framework for a description of the present-day complexity of the concept of the network enterprise. A number of definitions that I believe to be substantially innovative are put forward to give an account of this concept, carrying out in-depth investigations on its functioning and its connotations. The book presents a critical exploration of the evolution of production systems in the transition from the Tayloristic type of organization to that of the Japanese lean production models. The strategies involved in management of the network enterprise are examined, highlighting their ethical and cultural aspects, trailblazing a path that leads to rediscovery of the enterprise spirit and then analyzing the enterprise networks that are external to the network enterprise itself; such enterprise networks are seen as new phenomena capable of energizing small- and medium-sized production units.

And since this is a book based on the network model, it is my pleasure to conclude this "dedication" with a statement by Denis Diderot, once again taken from his *Thoughts on the Interpretation of Nature*, a statement that in many ways foreshadowed his conception of a world pervaded with hidden connections, and therefore interpretable as a network within which all things are linked. Thus wrote the French philosopher: "6. When one compares the infinite multitude of the phenomena of nature with the limits of our understanding and the frailness of our organs, can one ever expect the slow pace of our work, its prolonged and frequent interruptions, and the rarity of creative geniuses, to yield anything other than just a few shattered and disconnected shards of the great chain that connects all things?"

Of these thoughts I fully share both the form and the content.

Notes

1. Titus Lucretius Carus, *De Rerum Natura* [*The Nature of Things*], The Internet Classics Archive, http://classics.mit.edu/index.html. Translated by William Ellery Leonard.
2. Francesco Flora, *Introduzione a Galileo Galilei: Dialogo dei Massimi Sistemi*, Mondadori, Milan, 1996, xvi.
3. Denis Diderot, *Thoughts on the Interpretation of Nature*, Essay, 1753, http://www.chez.com/ottaviani/diderot/dpnature.htm. Fragment translated by Rachel Costa.
4. Denis Diderot, ibid.
5. Ibidem.

Bari Politecnico University Gianfranco Dioguardi

Series Foreword

The Springer Book Series on *Innovation, Technology, and Knowledge Management* was launched in March 2008 as a forum and intellectual, scholarly "podium" for global/local (gloCal), transdisciplinary, trans-sectoral, public–private, leading/"bleeding"-edge ideas, theories, and perspectives on these topics.

The book series is accompanied by the Springer *Journal of the Knowledge Economy* which was launched in 2009 with the same editorial leadership.

The series showcases provocative views that diverge from the current "conventional wisdom" that are properly grounded in theory and practice and that consider the concepts of ***robust competitiveness***,[1] ***sustainable entrepreneurship***,[2] and ***democratic capitalism***,[3] central to its philosophy and objectives. More specifically, the aim of this series is to highlight emerging research and practice at the dynamic intersection of these fields, where individuals, organizations, industries, regions, and nations are harnessing creativity and invention to achieve and sustain growth.

Books that are part of the series explore the impact of innovation at the "macro" (economies, markets), "meso" (industries, firms), and "micro" levels

---

[1]We define *sustainable entrepreneurship* as the creation of viable, profitable, and scalable firms. Such firms engender the formation of self-replicating and mutually enhancing innovation networks and knowledge clusters (innovation ecosystems), leading toward robust competitiveness (EG Carayannis, *International Journal of Innovation and Regional Development*, v. 1, n. 3, 2009, pp. 235–254).

[2]We understand *robust competitiveness* to be a state of economic being and becoming that avails systematic and defensible "unfair advantages" to the entities that are part of the economy. Such competitiveness is built on mutually complementary and reinforcing low-, medium-, and high-technology and public and private sector entities (government agencies, private firms, universities, and nongovernmental organizations) (EG Carayannis, *International Journal of Innovation and Regional Development*, v. 1, n. 3, 2009, pp. 235–254).

[3]The concepts of *robust competitiveness* and *sustainable entrepreneurship* are pillars of a regime that we call "*democratic capitalism*" (as opposed to "popular or casino capitalism"), in which real opportunities for education and economic prosperity are available to all, especially – but not only – younger people. These are the direct derivatives of a collection of top-down policies as well as bottom-up initiatives (including strong R&D policies and funding, but going beyond these to include the development of innovation networks and knowledge clusters across regions and sectors) (EG Carayannis and A. Kaloudis, *Japan Economic Currents*, January 2009, pp. 6–10).

(teams, individuals), drawing from such related disciplines as finance, organizational psychology, R&D, science policy, information systems, and strategy, with the underlying theme that in order for innovation to be useful it must involve the sharing and application of knowledge.

Some of the key anchoring concepts of the series are outlined in the figure below and the definitions that follow (all definitions are from EG Carayannis and DFJ Campbell, *International Journal of Technology Management*, 46, 3–4, 2009).

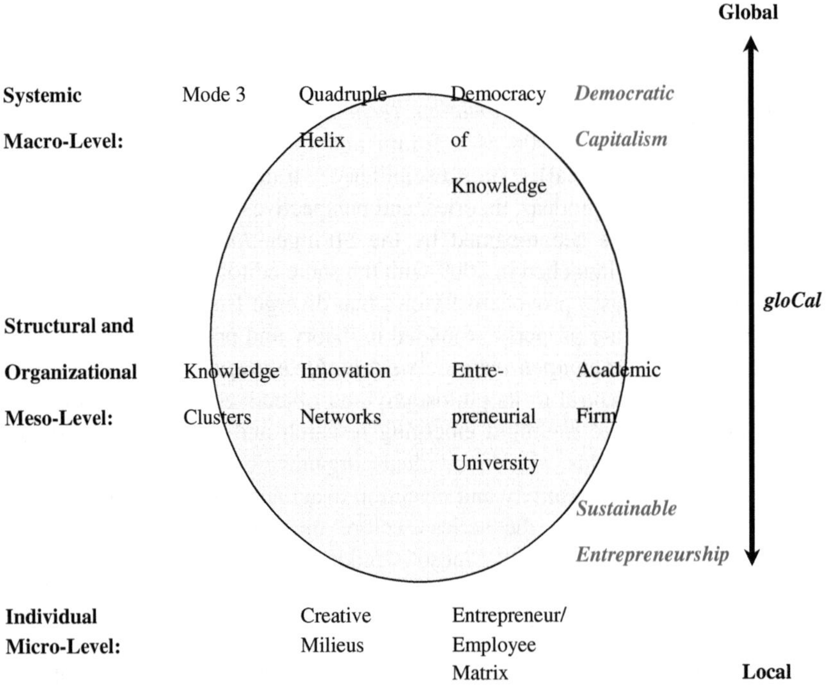

Conceptual profile of the Series on Innovation, Technology, and Knowledge Management

- The "Mode 3" Systems Approach for Knowledge Creation, Diffusion, and Use: "Mode 3" is a multilateral, multinodal, multimodal, and multilevel systems approach to the conceptualization, design, and management of real and virtual, "knowledge-stock" and "knowledge-flow," modalities that catalyze, accelerate, and support the creation, diffusion, sharing, absorption, and use of co-specialized knowledge assets. "Mode 3" is based on a system-theoretic perspective of socioeconomic, political, technological, and cultural trends and conditions that shape the co-evolution of knowledge with the "knowledge-based and knowledge-driven gloCal economy and society."

- Quadruple Helix: Quadruple Helix, in this context, means to add to the triple helix of Government, University, and Industry a "fourth helix" that we identify as the "media-based and culture-based public." This fourth helix associates with "media," "creative industries," "culture," "values," "lifestyles," "art," and perhaps also the notion of the "creative class."
- Innovation Networks: Innovation Networks are real and virtual infrastructures and infra-technologies that serve to nurture creativity, trigger invention, and catalyze innovation in a public and/or private domain context (for instance, Government–University–Industry Public–Private Research and Technology Development Co-opetitive Partnerships).
- Knowledge Clusters: Knowledge Clusters are agglomerations of co-specialized, mutually complementary, and reinforcing knowledge assets in the form of "knowledge stocks" and "knowledge flows" that exhibit self-organizing, learning-driven, dynamically adaptive competences and trends in the context of an open systems perspective.
- Twenty-First Century Innovation Ecosystem: A 21st Century Innovation Ecosystem is a multilevel, multimodal, multinodal, and multiagent system of systems. The constituent systems consist of innovation meta-networks (networks of innovation networks and knowledge clusters) and knowledge meta-clusters (clusters of innovation networks and knowledge clusters) as building blocks and organized in a self-referential or chaotic fractal knowledge and innovation architecture (Carayannis, 2001), which in turn constitute agglomerations of human, social, intellectual, and financial capital stocks and flows as well as cultural and technological artifacts and modalities, continually co-evolving, co-specializing, and co-opeting. These innovation networks and knowledge clusters also form, re-form, and dissolve within diverse institutional, political, technological, and socioeconomic domains including Government, University, Industry, Nongovernmental Organizations and involving Information and Communication Technologies, Biotechnologies, Advanced Materials, Nanotechnologies, and Next Generation Energy Technologies.

Who is this book series published for? – The book series addresses a diversity of audiences in different settings:

1. *Academic communities:* Academic communities worldwide represent a core group of readers. This follows from the theoretical/conceptual interest of the book series to influence academic discourses in the fields of knowledge, also carried by the claim of a certain saturation of academia with the current concepts and the postulate of a window of opportunity for new or at least additional concepts. Thus it represents a key challenge for the series to exercise a certain impact on discourses in academia. In principle, all academic communities that are interested in knowledge (knowledge and innovation) could be tackled by the book series. The interdisciplinary (transdisciplinary) nature of the book series underscores that the book series scope is not limited a priori to a specific basket

of disciplines. From a radical viewpoint, one could create the hypothesis that there is no discipline, where knowledge is of no importance.

2. *Decision makers* – private/academic entrepreneurs and public (governmental, sub-governmental) actors: Two different groups of decision makers are being addressed simultaneously: (1) private entrepreneurs (firms, commercial firms, academic firms) and academic entrepreneurs (universities) interested in optimizing knowledge management and in developing heterogeneously composed knowledge-based research networks and (2) public (governmental, sub-governmental) actors that are interested in optimizing and further developing their policies and policy strategies that target knowledge and innovation. One purpose of public knowledge and innovation policy is to enhance the performance and competitiveness of advanced economies.
3. *Decision makers in general:* Decision makers are systematically being supplied with crucial information, for how to optimize knowledge-referring and knowledge-enhancing decision-making. The nature of this "crucial information" is conceptual as well as empirical (case study based). Empirical information highlights practical examples and points toward practical solutions (perhaps remedies); conceptual information offers the advantage of further driving and further carrying tools of understanding. Different groups of addressed decision makers could be decision makers at private firms and multinational corporations, responsible for the knowledge portfolio of companies; knowledge and knowledge management consultants; globalization experts, focusing on the internationalization of R&D, S&T, and innovation; experts in university/business research networks; and political scientists, economists, and business professionals.
4. *Interested global readership:* Finally, the Springer book series addresses a whole global readership, composed of members who are generally interested in knowledge and innovation. The global readership could partially coincide with the communities, as being described above ("academic communities," "decision makers"), but could also refer to other constituencies and groups.

Elias G. Carayannis
Series Editor

Contents

About the Author

Gianfranco Dioguardi is professor of Engineering at the Politecnico University of Bari, where he teaches Corporate Economics and Organization. His extensive entrepreneurial experience, as CEO and COO of high-tech and construction companies, has endowed him with in-depth knowledge of the theory and practice of organizations.

He is the author of a wide range of articles and books (in Italian, French, and English) that address management and entrepreneurship, as well as the world of ideas, with particular reference to history, philosophy, and science.

Some of the topics presented in this book, such as the concept of "macrofirm," were originally introduced into economics by Dioguardi in 1983 in an ASCE publication. In 1996, the Italian Ministry of Education awarded him the gold medal for outstanding service to the School of Culture and Art, bestowed upon him by the President of the Italian Republic, and in 2001, upon the proposal of the Ministry of Universities and Scientific and Technological Research, he was awarded a similar medal for outstanding service to Science and Culture. Since 1989 he has been a Knight of the Italian Order of Merit for Labor and since 2004 a Knight of the Legion of Honor of the French Republic.

Chapter 1
Toward a Definition of the "Network Enterprise" Model

The profound and widespread revolution ushered in by the advent of advanced technology, above all the introduction of the personal computer and *Information and Communication Technologies (ICT)* in production industries and more generally in all operative institutions, has had far-reaching repercussions on corporate organization. An organizational model shaped by the network concept is becoming increasingly predominant. The descriptive term often used in reference to a network is that of a "Web" – the image is evocative of a spider's web – spreading out from a coordinating focal point which fulfils the function of corporate governance, with special emphasis on governance of production. To gain further insight into this model, it is helpful to start by examining the underlying nature of an enterprise and its operative functions, both epistemologically and in terms of corporate structure.

In a very general sense, the network enterprise is an organization composed of an information technology network (computers and robots), guided by a network of individuals whose mode of operation can be captured by the term "self-entrepreneurs." Such a phrase implies that they have the ability to perform their activity through a network of operative decisions which they themselves determine. Seen in the perspective of production, the network concept extends upstream to include a constellation of interconnected suppliers and allied industries which, taken together and in conjunction with the enterprise that stands at the focal point of the "web," make up what can be defined as a "macroenterprise." There is frequently an equivalent mirror image downstream, composed of a network of customers linked to one another by the organization of the "sales" function or more generally by corporate "marketing." Both the sales and the marketing functions are entrusted with coordination of the customer network through specific governance processes. This is in harmony with today's increasing awareness of responsibility toward the customer, whereby the customer is more and more interpreted as a veritable commissioning client. Thus even the private enterprise is taking on characteristics of a public nature, with all the social and ethical responsibilities this involves.

A detailed definition of the network enterprise calls for a series of clarifications to elucidate its contents and rationale. First and foremost, it should be underlined that the network model does not necessarily invalidate the more traditional function-based organization charts. Rather, it offers a complementary approach, illustrating and specifying the operative modes of the traditional format and reconsidering the

G. Dioguardi, *Network Enterprises*, Innovation, Technology, and Knowledge Management 104, DOI 10.1007/978-1-4419-1333-3_1, © Gianfranco Dioguardi, 2010

definition of an enterprise in the light of the present-day tendency for local aspects to be manifested in a global framework.

The entrepreneurial network can be seen as having a bipolar character, a feature that is common to all production enterprises. First, an enterprise is defined by a pole that centers on "general management," which has a predominantly strategic nature. It is here that decisions destined to shape the future of the enterprise are taken. This pole therefore displays the typical features of globality, with the need to take into consideration all the opportunities available on the global markets. It constitutes the area which, in Michael E. Porter's "chain of value," is set in the framework of infrastructural activities, human resource management, technology development, and sourcing.[1]

A second pole, which could be defined as a "production pole," concerns the production process itself. This is generally represented by the factory or the industrial plants rooted in the surrounding area. It thus has a typically local character. In Porter's above-cited "chain" this second pole corresponds roughly to the areas of incoming and outgoing logistics, operative activities, marketing, and sales.[2] The two poles constantly interact by virtue of the intricate network of relations between the strategic area and the production area.

The "general management" pole is characterized by a predominance of "interactive" relations arising among the various figures involved. One can in fact speak of a "hierarchy" that can be expressed through a network of cooperative working activities springing from the overall set of hierarchical interactions. The "production" pole, on the other hand, is characterized by a network of "transactions" which leads to an internal "quasi-market," although the latter is also open to the outside through its own web of suppliers and subcontractors. Thus in the case of the production pole one can speak of a typical transactional organization.

Taken as a whole, the enterprise can therefore be defined as a bipolar organization composed at one and the same time of a hierarchy and a transactional quasi-market. These two poles are interfaced by intense and constant interaction.

Within the production pole, the transactional network assumes a variety of forms in relation to the specific production process in question. It is manifested upstream as a network of suppliers whose products, services, and activities – often supplied *just in time* – enable the industrial plants to carry out the required production. Since the suppliers are effectively enterprises that contribute to achievement of the production goals of the overall central enterprise, they collectively make up the "macroenterprise," that is, the "system of enterprises" devoted to production.

The enterprise develops its production processes through a factory or a network of local production units. The units may be distributed geographically over an extensive area and may be situated at a notable distance from one another, in which case they can be examined in the framework of a global enterprise.

Factories are the typical places where innovation processes are devised and tried out. Frequently, an innovative concept springs from activity carried out within the general production pole, where innovation often arises as a response to the constant flow of new requirements expressed by customers. Increasingly, customers act as clients commissioning a product, and thus in order for the bipolar entrepreneurial network to achieve success it is vital to adopt a totally customer-oriented approach. This ensures that the customer is provided with a continually renewed range of product-linked services so as to ensure constant improvement of quality and performance.

It cannot be overemphasized that the quest for high levels of product quality and performance must be a priority issue in the drive toward entrepreneurial innovation. A product must be interpreted as akin to a process that can generate services characterized by excellence.

The fundamental core – a core with strategic connotations – from which the operative functions spread out represents a state function, composed of a hierarchy ordered according to specific functional organization charts that provide the services required for management. This picture implies a network of interpersonal relations manifested as interactions among the individuals involved, with exchange processes that tend to be immaterial as they refer almost exclusively to information and decisions. However, such processes also imply a web of assignments in terms of delegated responsibilities on which the subdivisions of the corresponding organization charts are based. In this framework, a company incurs period-based costs, that is to say, fixed costs, as compared to production costs. Thus the "state" concept stands in opposition to the "flow" that characterizes the production process.

The production flow is governed and guided by the general management core. It can be described as an operative network of transactions that unfurls through a flow of resources that have been acquired upstream; once these have been transformed by the production process, they give rise downstream to the products or services destined to the enterprise's customers. Therefore, an exchange market is present both upstream and downstream, its prices constituting "costs" upstream that vary directly according to the products to be produced, while downstream they represent the enterprise's revenue, which is likewise variable. It is important to underline the variability of both these factors – a variability that is dependent on the production – since it is precisely this aspect that differentiates them from the general fixed costs, which are typical of period-based costs, even though they accumulate over time. The flow that characterizes the production function is thus composed of a network of transactions which form the context for exchange of production factors.

In effect the enterprise is structured as a dual network, namely as an (interactional) network of (transactional) networks whose nodes are constituted by physical elements (types of technology, generally IT based) and by individuals or functional groupings of persons collectively entrusted with specific roles or operative functions. Overall, the enterprise presents a network structure consisting of a hierarchy (as is typical of international relations) and a market, the latter being characterized by exchange transactions.

This model, with its innovative features, supersedes the Tayloristic and Fordist organizational approach that characterized the firm and the industrial production plant prior to the introduction of IT, in particular the introduction of robotics. However, in the functional organization chart, coordination between the management and the roles needed to develop the different functions operatively still continues to be achieved through assignment of responsibilities and task delegation processes, which are accompanied by the chain of command necessary to ensure proper execution of the appointed tasks. This approach is then extended to governance of the actual management of the production function.

In the Tayloristic and Fordist scenario, task assignment was usually carried out according to a rigid breakdown of tasks, with the result that the assignee was also assigned the action to be carried out. The worker's operative freedom was thus severely restricted. All tasks were to be performed according to a mechanistic, and therefore rigid, approach. This organizational format was designed to establish a framework that could be likened to a clock, whereby each subject was to carry out his/her task in one way only, the way considered to be the *one best way*.

The rigid approach of Taylorism required the preliminary planning and splitting up of complex actions. This was arranged by purpose-established offices devoted to these specific functions, and it resulted in a tendency to simplify every activity to the utmost in order to facilitate execution of the required actions. It also had the effect of rendering actions repetitive in the context of global labor. This was the case not only for manual operations in the factory but for ordinary management (administrative) functions expressed by the organization charts. Each person in charge was constrained by a sort of decision-making passiveness determined by the preliminary planning processes. Such processes were particularly rigid and embodied the tool through which the individual lines of the corporate approach were delineated and defined.

The situation changed radically with the advent of IT, in particular with the widespread use of personal computers and robots. IT rapidly became commonplace in ordinary management – especially in business administration – and in production itself. Today an enterprise, and in general any operative organization, can be seen as a set of IT machines, computers, or related facilities, with varying degrees of interconnection, which are capable of carrying out ordinary management, with particular emphasis on the administrative aspects. But even production factories have also become the dominion of "robots," that is to say, of automated tools that can perform a large quantity of manual tasks and which are guided in their operative functions by dedicated computers.

Generally, all forms of IT are composed of a fixed piece of apparatus – the hardware – comparable to the structure of an industrial plant, its specific activities being defined and guided by purpose-designed programs – software – which have the characteristic of being easily modifiable. Software programs can thus introduce changes in real time and modify the activities produced by the fixed structures. This allows elevated operative flexibility. Computers and robots are governed by their designated operators, who enter into a singular one-to-one working relationship, almost an indissoluble bond, with the machines. To the observer, the enterprise thus appears as a complex operative context, where the unit of investigation is in fact this working pair that operates in a close-knit operative *rapport*.

It is indeed a most remarkable relationship that arises within these man–machine couples, which can be likened to a veritable assignment, a "technological assignment." In effect, the computer and its associated technologies no longer operate merely as prosthetic amplifications of human faculties. Rather, they exhibit something resembling a situation of independence made possible by a process of assignment of responsibilities, whereby the individual can command the execution of various actions which the machine then executes in a fully independent fashion, on the basis of its own programs, leaving the task of control of the final results to the operator. One may go so far as to suggest that when the computer is elaborating what it has been instructed to perform, a genuine operative autonomy takes shape, so that the machine becomes independent of the operator. This implies the existence of a period of time during which the will of the operator is in abeyance, giving rise to nothing short of a "principle of independence and non-determination of the computer." In such an interpretation, there can be no interaction with the machine while it is operating, unless it be by modifying the previously transmitted instructions and thus taking up a precise interval of time.[3]

In order to allow the relation of assigned responsibility to become established, it is necessary for the operator who is the protagonist of the event to be endowed with specific knowledge and background, which in turn implies a good educational level. Accordingly, today a person's educational background is playing an increasingly innovative role, one that needs to be developed in a more generalized manner in the context of the enterprise. This is one of the most momentous aspects of the IT revolution. Furthermore, it justifies the definition of the network of individuals as operators who are "self-entrepreneurs," in the sense that when liaising with the machine an individual must have a very clear perception of the decision-making scenario: deciding what tasks are to be assigned to the computer is not altogether dissimilar from the entrepreneurial decision-making process.

The possibility of easily modifying the performance of IT machines by simply substituting their software programs is in itself a major revolution, allowing the performances obtained from fixed industrial installations to be rendered flexible. The new frontiers of organization are thus oriented toward the search for global flexibility, not only in terms of general management of the enterprise but also as regards the production function in the factory. This phenomenon is becoming the distinctive characteristic of the enterprise today, but its beginnings can be traced back as far as the 1970s and 1980s.

Some interesting observations on flexibility are offered by Giovanni Azzone and Umberto Bertelè in their book entitled *L'impresa* [The Enterprise]: "*Flexibility* – understood as the ability and/or possibility for the enterprise to react rapidly and/or without excessive extra costs to changes in the context – generally tends to limit risk, by assuring that outcomes of work orders are less 'distant' from one another. That is to say, the outcomes are characterized by a lower degree of *variance*, for a 'spectrum' of possible alternative future scenarios. By contrast, *rigidity*, i.e. the presence of *constraints* that make it difficult for the enterprise to adapt to changes in context, tends to amplify risk."[4]

To facilitate the achievement of production process flexibility, it was suggested in the later decades of the 20th century that the production cycle should no longer be carried out entirely in-house. The so-called *make* approach, which was typical of the earlier processes and tended to render such processes rigid due to the presence of a multiplicity of fixed installations, was thus superseded by a more flexible form of organization. It now became feasible to buy in components from outside whenever these were required for the production of specialist parts. This was the *buy* logic, whereby a large proportion of the components of the final product were purchased from external enterprises, with the factory mainly carrying out assembly operations.

The distinction between these two modes of organization is aptly described by Azzone and Bertelè: "At one extreme, the enterprise can carry out in-house almost all of the activities necessary for the production of a product or a family of products, that is to say, it can have an elevated degree of *vertical integration.* At the other extreme, the enterprise can be strongly *deverticalized*, namely, it can limit its interest to the phases that are 'close to the market' (such as marketing, sales and after-sales care), and buy from external – *outsourced* – enterprises the physical goods it sells, the logistical services needed to deliver goods to customers and even the research and development services necessary for the extremely important phase of *product conception.*

Analogously to the observations on the business portfolio, the *portfolio of activities* the enterprise must have on its books for each business is not predetermined. Rather, the choice of 'what to make in-house and what to outsource' – commonly referred to as *make or buy* – represents a relevant component of the enterprise's strategy. This component has to be subjected to frequent review in order to ensure that the various opportunities which may arise are taken into account, and likewise the risks that may be incurred."[5]

The new enterprise implements its strategies of buying in from external sources by surrounding itself with a multiplicity of suppliers, which operate upstream from the production processes. These suppliers constitute the enterprise's "network of interconnected suppliers," giving rise to a systemic corporate conception which, taken as a whole, is able to achieve the final production. Thus what one has is effectively a "system of enterprises" defined as the "macroenterprise." The mirror image of this structure is found downstream, where product sales are generally distributed over a network of multiple customers. Such a configuration is characteristic, for example, of the type of enterprise – generally a service provider – known as *(Public) Utilities.*

According to Antonio Martelli, "The *utilities* (public utility enterprises) provide goods and services held to be of significant interest (designed to satisfy indispensable primary needs) for a community [...]. The sectors in which the supply of public services has developed most extensively are *water*, *energy* (electricity, gas, heating), *communications* (telephony, postal services), *transportation* (air and rail travel, urban transportation) and *urban hygiene* (waste disposal), as well as some sectors of relatively lesser importance such as *pharmacies* and *education* [...] The current European situation in the public service industry displays a heterogeneous assemblage of players, classified into *multi utilities*, enterprises that provide two or more public utility services [...]; *energy companies*, enterprises that provide only energy services (power and natural gas) to the final customers; *traditional utilities*, enterprises operating in only one public utility sector [...]."[6]

The *public utilities* industry is characterized by a number of structural elements, one fundamental aspect being its "network structure, from which there derives the definition of *network industries*, based on the necessary physical connection between the producing industry and the final user."[7] In this context "the model of the *multi-utility* properly speaking can be identified [. . .] in all corporate frameworks where there can be control (by shareholders or *de facto*) over a range of suppliers of utilities and in certain cases of other products or services."[8] Multi-utilities may achieve synergies in various sectors, above all in the technical, commercial, and financial sector, by operating with the so-called *cross selling* technique, that is to say, "sale of a packet of products to one and the same customer, within the same sales process. Through this technique, by placing a single type of order the customer purchases various different products that do not necessarily belong to the same family [. . .]."[9]

Previously, enterprises of this kind operated almost exclusively on monopoly markets, whereas today they are increasingly subject to privatization phenomena and consequently have to adjust their mode of operation to cope with the challenge of mounting competition. This further explains why the network structure that is typical of sales processes downstream from production is now being extended throughout the organization of the enterprise, in order to achieve greater effectiveness and efficiency on the markets where the enterprise is present.

It is clear from the above observations that the network concept applied to the corporate world is currently undergoing a phase of notable development. Interestingly, this is giving rise to an image of continuity between the network enterprise and the enterprise networks which, at the local level, constitute the phenomenon of "local economic systems with elevated integration," better known as "industrial districts." In Italy the district phenomenon has been particularly successful.

Using a traditional description, the enterprise can be portrayed with the image suggested by Azzone and Bertelè: "The *enterprise* represents a component of a highly structured and extremely complex system – which we can term an *economic-financial system* – and it operates in close interdependence with other components within the overarching framework of this system.

The *inputs* it uses are goods and services from other enterprises within the system, such as raw materials and semifinished goods, energy, transportation services (for goods and persons), consultancy services. It sells its *output* – consisting of an often fairly diversified portfolio of goods and/or services (henceforth referred to as *products*) to the final consumers or other enterprises or the state administration. It finances its activities with the money from its shareholders (these may be private individuals, investment funds, pension funds, etc.), supporting its financial requirements via the stock market if it is listed on the stock exchange, or through loans granted mainly by banks."[10] This definition clearly underlines that there is generally a strong degree of integration between the enterprise and its environmental context, as well as a tendency for the enterprise to adopt a network form of operational model.

Among the authors who have conducted in-depth studies of this topic, mention should be made of Federico Butera, who has addressed the issue in a number of his books.[11] One particularly significant definition is contained in his observation that "[In the network enterprise] external enterprises and internal organizational units represent not only alternative suppliers selectable on the basis of costs and reliability, but also nodes or viable systems capable of their own autonomous strategy and endowed with structures that are adequate for development (genuine enterprises) [...]. We will define the organizational network (a concept that includes both the enterprise network and the network enterprise) as a system of recognizable multiple connections and of structures within which there are 'nodes' that have an elevated level of self-regulation (or viable open systems). Such nodes are capable of cooperating (i.e. of carrying out various types of effective transactions) in conceiving and conducting processes, as a means of working towards shared and/or compatible aims and cultural approaches. This system is energized, and to some extent designed and monitored, by a core organization ('the strategic agency'). Thus whether one considers the enterprise network or the network enterprise, in either case the old anarchical structure of the local market is being deconstructed and, with it, the old hierarchical structure of the large enterprise as well. These structures are being supplanted by a *new flexible enterprise structure*, based not on shareholdings or organization charts but rather on the recognizable results of economic value and visibility, governable processes, independent and holographic 'nodes,' governed connections and specific operative properties. This leads to composite structures that include hierarchies, markets, telematic systems, cultures and institutional cooperation systems, all coexisting at one and the same time. Basically, then, the network enterprise enshrines full permeability not only between enterprises and internal organizational units, and between the market and the organization, but also between enterprises, institutions and society."[12]

Starting out from these general considerations, Butera highlights a number of specific aspects. Of particular interest are those concerning the way the network enterprise is designed. It is worth taking another look at Butera's crucial concept of nodes as a structural element of the network: "From the point of view of the designing of the enterprise system, we distinguish between the '*natural network enterprise*' and the '*governed network enterprise.*' The natural network enterprise can be defined as the system of recognizable and multiple connections and structures within which there are 'nodes' that have an elevated level of self-regulation (viable open systems): such nodes are capable of cooperating in conceiving and conducting processes (i.e. in carrying out various types of effective transactions) so as to work towards common goals or shared aims. According to this definition, the network enterprise has the nature of an entrepreneurial subject and can be characterized as an organizational system, even in cases where it cannot be defined in terms of a legal identity (corporate name, corporate structure, statutes, etc.) or a unified hierarchical structure (organization charts, powers, assignments, corporate procedures, etc.). The network enterprise exists as an economic subject and as an organizational system complete with its own goals and control systems [...]."[13]

Elsewhere, Butera further states "When individual or collective, public or private entrepreneurial subjects, each endowed with an identity of their own, deliberately proceed to design, manage and in the overall sense maintain a system such as the one described, [. . .], then what we have is a governed network enterprise. As in any other type of enterprise, one of its essential characteristics is the need for identification of the ends and control over the results. But in the governed network enterprise one does not necessarily find a single focal point where plans and control are effected. Governance of the governed network enterprise is a collective process which is designed – more or less consciously – to ensure viability of the nodes, select and develop the critical connections, set up forms enabling multiple structures to coexist, and develop operative mechanisms. There may or may not be a formal organism (a leading enterprise) or an informal arrangement (a governing committee) in which this process is crystalized and translated into the final decision. But it is the players themselves who breathe life into the network [. . .]."[14]

A document issued by FIAT, mentioned by Enrico Auteri, proposes a useful definition on this subject: "The *network enterprise* is a *network of subjects endowed with excellence* in their own field and oriented towards a common objective. Their significant characteristics are as follows: – what binds them together is that the aim of optimizing their reciprocal skills; – they make strategic use of relations with their partners; – they obtain a 'return' that can be measured in economic terms. It is a kind of enterprise in which the concept of the *boundary of the firm* becomes increasingly evanescent [. . .]. It is composed of nodes and connections: the internal and external organizational and/or corporate structures can be described as self-contained *nodes* of the network, capable of enhancing their own resources and skills and of keeping up their level of competitiveness over time; at the same time, the *nodes* are linked to one another and with the central enterprise by legal, economic, financial and organizational *connections* that express relations oriented towards optimization of their reciprocal skills. This interlocking configuration makes it possible to achieve a value that is measurable, both for the global economy of the individual enterprises of the Group, and also for the Group taken as a whole."[15]

Auteri provides a further illustration to clarify the entrepreneurial revolution that has led to the concept of network enterprise: "The 1992/1993 crisis heightened the tendency to move towards a type of enterprise that would be capable of reducing non indispensable coordination costs, and would at the same time devise models for transferring decision-making costs, activities and responsibilities to the outside. This ushered in an intensification of outsourcing and tertiarization operations. The ensuing organizational approach then gave rise to what became known as the 'transactional company,' although such an organizational system still remained rather distant from the 'network enterprise,' That is to say, although in popular terminology the network enterprise and the transactional company were often considered synonymous, the network enterprise ultimately represented the richest and most complex form of this evolutionary trend. This was partly because the network enterprise was characterized, among other things, by elevated flexibility in its functioning, especially in times of ever increasing turbulence, and by greater complexity of management."[16]

In addition to comments on the rise of the transactional enterprise, the author also points out that "[...] [t]here was growing awareness of the need for an evolution. What was called for, on the one hand, was a form of enterprise that would be capable of reducing non necessary coordination costs, namely both hard and soft costs (a short enterprise that would reduce hierarchical levels and structural expenses, while at the same time increasing the transverse nature of its functioning). But there was also a need for the development of models that would allow costs and activities to be transferred towards the outside. The new approach resulted in an increasing number of cases of externalization (divestment of operative activities to outsourced – but also sometimes to internal – enterprises), as well as tertiarization operations (divestment of service activities). Such moves made it easier to increase structural relations with third parties, i.e. with suppliers and dealers. It was this organizational tendency that became known as the 'transactional enterprise.' However, it was not destined to represent the final arrival point of the evolutionary trend, for the mere fact of outsourcing and tertiarizing was by no means sufficient to create a new organizational system. What was also required was to provide an accurate statement of the relations and to establish who the new system was to be built with, and how it was to be designed, independently of the legal boundaries of the enterprise."[17]

The evolutionary trend thus ultimately led to the network enterprise. Today, in addition to its flexibility, one of the pervasive characteristics of the network enterprise consists of the services provided by its corporate functions, both within the enterprise itself and also on the outside in the form of services to customers using its products-services. Thus the network enterprise effectively takes on the characteristics of a strongly tertiarized organizational structure.

Notes

1. Michael E. Porter, *Il vantaggio competitivo* [*Competitive Advantage*, The Free Press, New York, 1985]. Ital. Transl. Edizioni Comunità, Milan, 1987, p. 46.
2. Ibidem.
3. Gianfranco Dioguardi, *L'impresa nell'era del computer* [*The Enterprise in the Era of the Computer*], Edizioni Il Sole 24 Ore, Milan, 1986, and Gianfranco Dioguardi, *I sistemi organizzativi* [*Organizational Systems*], Bruno Mondadori, Milan, 2005.
4. Giovanni Azzone, Umberto Bertelè, *L'impresa: sistemi di governo, valutazione e controllo* [*The Enterprise: Systems of Governance, Evaluation and Control*], Second edition, Etas, Milan, 2003, p. 84.
5. Ibidem, p. 59.
6. Antonio Martelli, *Strategie di «cross selling»* ["Cross Selling Strategies"] in *Sviluppo & Organizzazione* [*Development and Organization*], n° 207, January–February 2005, pp. 19–20.
7. Alberto Dell'Acqua, *Public utilities e dinamiche evolutive, Public Utilities, creazione del valore aggiunto e nuove strategie* [*Public Utilities: Added Value Creation and New Strategies*], Egea, Milan, 2001, p. 8.
8. Alberto Dell'Acqua, Gianfranco Gianfrate, in *Public Utilities...* cit., p. 161.
9. Antonio Martelli *Strategie di «cross selling»* ["Cross Selling Strategies"], cit., p. 27.
10. Azzone, Bertelè, *L'impresa...*[*The Enterprise...*], cit., p. 53.

11. Federico Butera, *Il castello e la rete. Impresa, organizzazioni e professioni nell'Europa degli anni '90* [*The Castle and the Network: Enterprise, Organizations and Professions in 1990s Europe*], Angeli, Milan, 1990, and also Federico Butera, *Il campanile e la rete. L'electronic business e le piccole e medie imprese in Italia* [*The Belltower and the Network: Electronic Business and Small and Medium-Sized Enterprises in Italy*], Edizioni Il Sole 24 Ore, Milan, 2001.
12. Federico Butera, *I cambiamenti nella qualità dell'occupazione nell'era delle tecnologie dell'informazione e della comunicazione: nuove forme di organizzazione e di professioni* ["Changes in the Quality of Employment in the Information Technology Era: New Forms of Organization and Professions"], in *Atti dei Convegni dei Lincei* [*Proceedings of the Lincei Academy Conferences*], n° 139, Accademia Nazionale dei Lincei, Rome, 1998, p. 379.
13. Federico Butera, *Il castello e la rete* [*The Castle and the Network*], cit., pp. 60–61.
14. Federico Butera, *Il campanile e la rete* [*The Belltower and the Network*], cit., pp. 81–82.
15. Enrico Auteri, *Dalla gerarchia alla leadership* [*From Hierarchy to Leadership*], Guerini e Associati, Milan, 2004, p. 433–434.
16. Enrico Auteri, *Dalla gerarchia alla leadership* [*From Hierarchy to Leadership*], cit., p. 301.
17. Ibidem, p. 304.

Chapter 2
The Complex Nature of the Network Enterprise

One of the major issues that is still of great current relevance is the inquiry into the real nature of the production enterprise, when such an enterprise operates in an economic system coordinated and governed by the price mechanism.

In 1937, Ronald H. Coase, an American economist of British origin, born in London in 1910 and Nobel Prize winner for economics in 1991, was the first to reopen the question addressed in an earlier age by the classical economists, concerning the nature of the enterprise and its manner of functioning on the market. Coase, who taught at the University of Chicago, published only a small number of works, but at least two seminal publications had a profound impact on economics. These were *The Nature of the Firm* (1937), written when he was 27, and *The Problem of Social Cost* (1960), published 23 years later.[1] In the first article, in particular, he set himself the task of attempting "to discover why a firm emerges at all in a specialized exchange economy." This ushered in the concept of negotiation costs (that is to say, transaction costs).

Such costs are crucial for the existence of the firm, in that firms can be said to "replace" a price system through the action of entrepreneurs, for whom being "their own master" as independent traders is no longer advantageous once use of the open market becomes more expensive than utilization of their own entrepreneurial organization.[2] Moreover, in his discussion on the nature of the firm, which Coase defines as "the system of relationships which comes into existence when the direction of resources is dependent on the entrepreneur," and which he further explains by stating that "When the direction of resources (within the limits of the contract) becomes dependent on the buyer in this way, that relationship which I term a 'firm' may be obtained."[3] Thus in its relation with the seller from whom it purchases its resources, the firm exists inasmuch as it offers a cheaper alternative to the price systems operating on the market.

Yet the article has been "much cited and little used," as Coase himself realized.[4] Among the many citations, one may quote the mention which appeared in *The Economic Nature of The Firm – A Reader*, which is an anthology where Coase is ranked not far below historically important contributions – from Adam Smith to Karl Marx, from Frank Knight to Friedrich Hayek and Alfred Chandler.[5] The citation in the anthology underlines the importance of "hypotheses" in economic theory and eloquently testifies to emblematic nature of Coase's statement that "[e]conomists in

G. Dioguardi, *Network Enterprises*, Innovation, Technology, and Knowledge Management 104, DOI 10.1007/978-1-4419-1333-3_2, © Gianfranco Dioguardi, 2010

building up a theory have often omitted to examine the foundations on which it was erected."[6]

The scholar who has most closely followed the path laid out by Coase is certainly Oliver E. Williamson, born in 1932 in Wisconsin and holder of a professorship at the University of California at Berkeley. Among his most important books the celebrated work *Market and Hierarchies: Analysis and Antitrust Implications* should be mentioned and likewise *The Economic Institution of Capitalism: Firms, Markets, Relation Contracting,* as well as *The Mechanisms of Governance*, all of which were also published in Italy.[7] In addition, Williamson is the editor and coeditor of numerous successful anthologies such as *The Firm as a Nexus of Treaties.*[8]

Romano and Danese offer the following introduction to the topic: "Transaction theory, or transaction cost economics, which was initially formulated by Williamson (1975, 1985), represented one of the central issues of organizational debate starting from the mid 1970s. Unlike in neoclassical economics, the elementary unit of analysis of transactions is no longer the good that is produced, but rather the transaction itself, understood as the transfer of a good or a service through a technically separable interface, which involves an exchange of values between the parties."[9]

Williamson proposed that the immediate unit of analysis should be the concept of "transaction." In *The Mechanisms of Governance* he stated that "The transaction is the basic unit of analysis, whereas orthodoxy is concerned with composite goods and services. Pin making, how to organize more generally, how to govern the eighteen distinct operations (transactions) made famous by Adam Smith (1776) rather than how many pins to make and at what price, becomes the object of the analysis."[10] And with regard to the "structure of governance," what particularly interested him was that "Whereas the neoclassical firm is defined as a production function (a technological construction), transaction cost economics describes the firm as a governance structure (an organizational construction). According to the first definition, the firm is a technological black box in which inputs are transformed into outputs without reference to the organization. According to the second definition, firms and markets are alternative modes of governance (Coase 1937), and the allocation of activity between firms and markets is not taken as given but is something to be derived."[11]

However, the transaction per se cannot be considered, in general, as the only unit of investigation. In today's complex entrepreneurial world a new framework of reference is emerging: the pair consisting of the individual and the computer, which operates in unison within the corporate context. But within the environmental scenario, i.e., within the market framework, it is also possible to take the enterprise itself, in its complex composition, as the unit of investigation. In this case, the enterprise is considered as an exchange node, i.e., as a routing node for transactions taking place upstream and downstream from its management, with the predominant interpretation of the "production" function, as it is usually defined by classical economy.

On the market, however, the enterprise continues to maintain the characteristics indicated in its definition given in Chapter 1 as a bipolar network, i.e., as the expression of a general pole (hierarchy), which performs its own specific activities, and a typically productive (transactional) pole which is expressed through the characteristics of the classical "production function." Thus according to this definition, the institution of entrepreneurship is not merely a production function, as maintained by neoclassical economics: rather, it is a set of hierarchies and markets which form the setting for transactions. "Hierarchy" is a crucial concept here, as it is the hierarchy that gives rise to the entrepreneurial structure; moreover, through hierarchy it becomes possible to ensure the *governance* of production transactions by means of contracts that form the basis for such transactions.

The concept of entrepreneurial organization thus takes on a far broader significance. Rather than merely suggesting a hierarchical bureaucracy, the perspective expands to include the concept of the market with its intermediate forms of "quasi-market," thereby also introducing the context of contracts.

Market transactions, which Williamson proposes to take as the unit of investigation, display characteristics of material transfer occurring upstream of the production function. Such transfers occur above all when the reference enterprise purchases from its suppliers the various goods, performances, services it needs to produce the products which will, in turn, be transferred in transactions destined to take place downstream, in product allocation processes.

A transaction is therefore characterized by economic transfers defined on the basis of a market price, and this is identified through a demand value and a supply value, reciprocally accepted by the seller and the purchaser. The price agreed upon may not necessarily be the lowest that can be obtained on the market, if the seller happens to offer competitive advantages capable of rendering the seller's product or services more attractive and therefore preferable as compared to other proposals.[12]

In this sense, transactions are substantially distinct from the "interactions" that develop as typical immaterial phenomena in the framework of the hierarchy. Interactions consist of relations that have grown up among the protagonists of corporate processes who habitually work together and engage in contact with one another. Therefore, interactions arise among individuals who are linked by relations of communication and information transfer. Such relations may also be manifested in terms of assignments of responsibilities and through forms of consultancy. What is generally lacking in these interactions is precisely the question of price as an exchange mechanism, in contrast to the typical price-governed nature of transactions.

The natural conclusion of the production process comes about with the transfer of goods and services to customers, with the additional possibility of product-related services regulated by contracts stipulated in advance. In this perspective the enterprise can be interpreted, with regard to its production function, as a node of transactions that are manifested on the market and which to some extent determine it.

Transactions involve the exchange of goods and services as well as the performance of various actions. As such, they have a specific economic nature, which translates into a characteristic of a financial type, defined in the framework of the completed transfer. The financial aspect is determined through the positions of debit and credit bound up with the transfer; this then leads to settlement of the payment, in such a manner that the transaction can also be characterized from a monetary point of view. The set of these connotations defines the transactions both upstream and downstream from the reference enterprise, in the typical market scenario in which this enterprise carries out its production function. It is through the production function that the production management of the enterprise, and the consequent economic outcome, is accomplished.

While admitting of a unitary interpretation, the enterprise presents a hybrid nature composed of a set of interactional and transactional relations that are manifested above all through contracts. Therefore, enterprises are not completely a market entity, nor indeed are they merely a phenomenon of hierarchy. Rather, they consistently present mixed characteristics, partly also as a function of what they achieve internally (the *make* logic awards priority to hierarchy) and of what they buy in from outside (the *buy* logic refers predominantly to the market).[13]

Special attention should be paid to the concept of the boundary of the enterprise, i.e., of an internal environment that is distinct from what is outside. As a first step toward explaining this conception, it may be said that the boundary coincides with entry barriers that are capable of subdividing the fundamental assets, namely human resources and technology. Given the fundamental assets, the hierarchical interactions will then develop by means of flow acquisitions (such as human resources, production resources, semi-finished products, various activities) that are typical of the production process which forms the object of the transactions and of the "quasi-market" characterizing that particular aspect of the enterprise.

The concept of the enterprise and the entrepreneur has also been examined by Alfred D. Chandler, above all in his important book on *Strategy and Structure*. According to Chandler, "[...] the term industrial enterprise needs to be defined. Used in a broad sense, it means here a large private, profit-oriented business firm involved in the handling of goods in some or all of the successive industrial processes from the procurement of the raw material to the sale to the ultimate customer. [...] An industrial enterprise is thus a subspecies of what Werner Sombart has described as the capitalist enterprise, which as an independent economic organism is created over and above the individuals who constitute it. This entity appears then as the agent in each of these transactions and leads, as it were, a life of its own, which often exceeds in length that of its human members.

While the enterprise may have a life of its own, its present health and future growth surely depend on the individuals who guide its activities. Just what, then, are the functions of the executives responsible for the fortunes of the enterprise?"[14] On this point, Chandler specifies that "The executives who actually allocate available resources are then the key men in any enterprise. Because of their critical role in the modern economy, they will be defined in this study as entrepreneurs. In contrast, those who coordinate, appraise and plan within the means allocated to them will be termed managers. So entrepreneurial decisions and actions will refer to those which affect the allocation or reallocation of resources for the enterprise as a whole, and operating decisions and actions will refer to those which are carried out by using the resources already allocated."[15]

Chandler thus addresses the question of entrepreneurial strategy and structure, stating that "It seems wise here to emphasize the distinction between the formulation of policies and procedures and their implementation. The formulation of policies and procedures can be defined as either strategic or tactical. *Strategic* decisions are concerned with the long-term health of the enterprise. *Tactical* decisions deal more with the day-to-day activities necessary for efficient and smooth operations."[16]

The above-described aspects, together with the points mentioned by Chandler concerning the protagonists of the entrepreneurial phenomenon, highlight a further characteristic inherent in the nature of the enterprise: its intrinsic complexity capable of influencing not just its spirit but often even its behavior.

In effect, production enterprises can be seen as typical complex organizations having a structure that is constantly in an evolutionary flux, forever alternating between regularity, complexity, and crisis. Precisely for this reason they need the ability to constantly express organizational innovation, conceived as an eternally new frontier to be conquered.

Interesting observations on this situation are offered by Williamson, who in his book on *The Mechanisms of Governance* states that "[...] economic organization is made vastly more complicated by the numerous hazards that accrue to the combination of uncertainty with bounded rationality and opportunism."[17] An enterprise is in fact a typical example of a complicated system, but one that is constantly tending toward organizational complexity. The institution, taken as a network of individuals who have to be entrusted with sufficiently broad assignment of responsibilities to allow them significant operative independence, presents the risk of forms of self-organization on the periphery, manifested by subjects who participate in the entrepreneurial activity. This kind of risk should be considered as a specific manifestation of potential critical states of open systems, by analogy with the manifestations that can be found in the physical world, as noted in particular by Ilya Prigogine.[18]

Furthermore, the enterprise in its current network format avails itself of sophisticated IT facilities for its administrative management, which may end up by generating a sort of "technological determinism." This term has been used in a significant manner by Federico Butera, one of the greatest and most insightful scholars of complex organizations. Butera observes that "With telematic communication systems, information highways and the Internet, once again we are confronted with the 'technological determinism' which for over a century has been a recurrent feature of technological development, like an endemic plague. Technological determinism starts out from the assumption that each technological breakthrough can engender one society alone – indeed, that a breakthrough takes on and generates the fundamental characters of the society in which it arose and developed (the machine society, the automation society, the automobile society, the IT society, the telecommunications society, now even the Internet society etc.)."[19] A process is thus triggered with regard to which Butera comments that "[. . .] instead of a project that seeks to set up a society, an enterprise, a factory, a public service, there emerges the idea of technology that brings salvation (or threatens doom), promising to obtain within a short lapse of time what complex processes of change management struggle to represent and accomplish. It is an 'easy revolution' which comes devoid of any relation either to history or people."[20]

Basically, then, the configurations of hardware in the technologies used in enterprises generate a sort of determinism, which has been called "technological" determinism precisely because its source lies in technology. It is constantly influenced by the finalism which is typical of software that is managed by individuals going about their function of corporate operators. In fact Butera himself argues that "With regard to the development of production technologies, on the other hand, it was already demonstrated earlier (Butera 1989) that technology certainly has a revolutionary effect inasmuch as it 'disrupts' the constitutive order of the elements of production activities and of society itself: every time there has been a technological leap, this has either replaced elementary (production or consumption) activities, or modified coordination and control systems, or redesigned modes of communication and so forth."[21]

On closer inspection, however, it becomes clear that these technological effects are not produced by the machines in themselves, but rather by the way they are put to use through implementation programs (software). Since such programs can fairly easily be changed, they may also revolutionize the current state of affairs.

It is precisely this revolutionary potential of technology (and above all of information technology) that makes it imperative for an enterprise to ensure that its operators are strongly involved in the activities of the enterprise, so that the operative capabilities of the machines (computers) can be fully exploited. Thus the operators are called upon to repeatedly make adaptations in the emerging causality that is typical of the determinism-governed world of technology, in order to ensure constant alignment to the ends pursued by the organization. Such adjustments are introduced by dint of special programs or types of software which allow practical operative management of the machines. Therefore, the enterprise in turn undergoes continual adaptation to the external environment, according to a process that can be likened to the Lamarckian conception of evolution. In particular, what one finds is a multiplicity of component parts, interactions, and retroactions, self-organization phenomena – in other words, a situation of never-ending evolution, as has been the case throughout the history of the enterprise. This state of affairs ensues from an exchange, both of knowledge and also of concrete economic elements, between the internal environment and an external scenario that is by its very nature turbulent and consequently difficult to analyze and to map out according to precise plans.

That difficulties inevitably arise in the transition from an organizational "structure" to a "strategy" is clearly revealed by a glance at the case of the enterprise. An organizational structure is intrinsically "complicated," as are the organizations portrayed by Williamson, while a "strategy" is inherently complex. This issue has been addressed by another American scholar, Larry E. Greiner, who notes that "[. . .] Alfred Chandler Jr. in his book *Strategy and Structure* [...], argues that the prospects offered by the market determine corporate strategy, whereupon the company then determines its organizational structure. [. . .] In actual fact, this structure can have a decisive role in influencing corporate strategy."[22]

Insight can be gained relatively easily into the issue of structural complication through the study of organization charts. By contrast, it is far more difficult to follow and explain the evolutionary dynamics of complexity, which comes into play as a result of concrete activity by the structure designed to render the corporate strategy operative. One possibility open to analysts comes from an exploration of corporate strategy and planning. However, in the attempt to set up explanatory paradigms that will provide an account of complexity in the socioeconomic world, it is indispensable first to define a climate of operative "normalcy" characterized by a sufficient evolutionary stability of the history of the enterprise.[23]

When the threshold of normalcy is overstepped, a state of "crisis" is triggered, leading to a degree of complexity that is even more difficult to interpret and clarify.[24] In fact, in the lifetime of an enterprise it is by no means infrequent to find critical situations giving rise to a form of disorganized complexity that can be likened to chaotic processes.

It may be interesting, at this juncture, to reflect on the approach taken by scholars of complexity in examining the theme of the enterprise. For example, in the view of Joël de Rosnay, "the enterprise has a life of its own. It comes into being, grows, develops, reaches maturity and dies. Every enterprise is a production cell of the social organism [. . .;] an enterprise is any activity that sets itself the aim of selling a product or a service on the market of consumption or production commodities. [. . .] The enterprise performs [. . .] two main functions, one on the level of individuals, the other on the level of society. The first function is the production of the commodities and services that are of use in satisfying man's needs. The second consists in wealth creation, i.e. in generating, through its growth, a surplus of value [. . .]."[25]

Thus the enterprise, in all its manifestations, represents a veritable "socioeconomic laboratory" that can be of use in studying social complexity. Today the enterprise is increasingly shifting toward a governance structure, and thus it is becoming a governance structure for organizational coordination.

The organizational model of the enterprise oscillates between the hierarchy expressed by the complicated structure – which belies its name and is a bringer of order, as can be seen from its organization charts – and the market of internal and external transactions. In its guise as a transactional market, the enterprise displays a number of situations that have every appearance of being guided, as it were, by a sort of "invisible hand," linked to that specific internal market. The developments of such situations should be encouraged as far as possible, yet they should also be countered whenever this is deemed necessary. Such a move vis-à-vis the external world gives rise to what Chandler defines as the "visible hand" of managers.

With regard to the entrepreneurial institution, it is important to seek to safeguard tradition, inasmuch as it embodies the history of the enterprise. It is through awareness of this history that a better understanding of the state of the art of the system can be gained; this in turn facilitates management of complexity. Yet at the same time, precisely the concrete fascination exerted by tradition has the power to spark the emergence of innovation that is indispensable to propel the life of the enterprise into a dimension where the challenges of the future can be met.[26]

Once again, then, in order to interpret corporate complexity it is helpful to turn to the model of "network enterprise," through which the enterprise is portrayed as a set of "networks" capable of coordinating a multiplicity of phenomena.

Discussing the concept of network involves broaching the question of "order," which is indispensable for an in-depth understanding of complexity. In this perspective, attention generally focuses on a network of individuals, a network of interactions between operative poles, a technology network, and a network of other enterprises forming part of the suppliers and allied industries which, taken together, make up the macroenterprise.[27] This subject is brought up in a work by an English author, Ralph D. Stacey, who suggests that the network structure could also be represented as a functional – albeit highly decentralized – organizational structure or it could be represented by a complex structure based on a matrix. More specifically, one may view "complex matrix or network type structures as a means of bringing people together across functional and business unit boundaries."[28]

In the network enterprise, individuals need to be assigned a fairly broad range operative of tasks, going beyond the Tayloristic framework that was limited to the specific actions to be performed. It is therefore an assignment of responsibility that focuses on outcomes, so that each operator can devise the most appropriate manner to achieve the proposed goal.[29] This partially explains why it is relevant to approach the issue in terms of a network enterprise composed of veritable "self-entrepreneurs," who should be driven by what Stacey calls "entrepreneurial enthusiasm" capable of making persons "creative," in such a manner that they "will seek to interact creatively with the other people who constitute the environment of their business."[30]

Seen from this perspective, the most evocative image of the network enterprise is that of an intricate web of interconnected networks, subtly but deftly coordinated with one another in a vision that interprets the overall complex corporate unit.[31] The complex thereby created is a veritable "architecture of networks" which expresses the structure of the enterprise and acts as the hub from which the spokes of strategic and functional behavior radiate outward.

A first basic network arises between the poles consisting of the shareholders, the organs delegated by the shareholders for administration of the enterprise, and the genuine organizational structure – i.e., the management. Even at this first level, it is not infrequent to find interactions that may move in conflicting directions, leading to an effect of complexity. There can be difficulty, at shareholders' assemblies, in interpreting the various different goals and desires of the shareholders, especially if the body of shareholders is composed of a multiplicity of small shareholders devoid of a stable core group. The resulting divergences and disagreements immediately have repercussions on the administrative structures (board of directors and other governing bodies: president, managing directors); consequently, such conflicts also impact on those who are called upon to be the driving entrepreneurial force (the typical figure and function of the entrepreneur) and whose mission is to achieve the statutory objectives laid down by their contractual position within the enterprise.

This initial state of complexity can degenerate into a managerial – and more generally, operative – situation of extreme confusion, which has a highly adverse effect on the performance of the enterprise. Furthermore, the initial complexity has basic internal, i.e., structural, characteristics which cannot easily be solved through external intervention. And since complexity tends to appear in a random fashion, it may be difficult to adopt rational measures or antidotes to straighten out the situation.

In its managerial structure and organizational setup (organization chart), the enterprise can be seen as a network of individuals who carry out certain operations or functions designed to implement the corporate mission. Within this network conception, the hierarchical structure tends to flatten and become more horizontal, while continuing to be the expression of an order from which corporate operative intentions take shape. As Stacey points out in this regard, "All decision-making processes take place in a hierarchical structure. With highly complex environments, organizational balance can be maintained only by virtue of hierarchy."[32] Thus the network is manifested in a bipolar form, namely in terms of strategic hierarchy and operative production.

Development of corporate activities engenders fertile interaction among the various functions, thereby shaping a complex network of interrelated aspects which can be fully comprehended only if an organizational coordination function is put in place. This can more easily be achieved if it is supported by strategic corporate development plans that can be transformed into short-, medium-, and long-term planning and control tools (planning of a technical nature and administrative or budget tools).

The goal of further simplification of the complex system of corporate management, in a context where the enterprise is aiming to move toward future development, can at times be grasped more clearly by bearing in mind several types of behavior that are characteristic of the corporate environment. In his book *La vita imperfetta*, Furio Colombo outlines two typical modes of behavior: "[...] as in monastic orders, divided as they are into apostolic (those who operate) and contemplative (those who meditate) figures [...]."[33] Effectively, corporate human resources can be divided into two distinct types: those with entrepreneurial aptitudes and those with a natural predisposition toward more explicitly administrative functions.

The approach to problem-solving observed in the former of these two groups tends to be that of setting up the procedures necessary to achieve the established objectives. Therefore, this is a forward-looking approach, and the actions of this group have a more finalistic than deterministic connotation. The functions dealt with are generally linked to the world of production, marketing, or sales.

The administrative approach, on the other hand, goes hand in hand with awareness of the past, recording the history of the enterprise and situations that have already occurred. This group is certainly highly important, but its approach is less constructive and less oriented toward the future. Therefore, it may at times tend to hinder entrepreneurial development. The typical functions in this area concern the administrative sector and corporate control. Generally, administrative figures perform their role with a marked degree of pessimism. This deleterious propensity toward pessimism needs to be mitigated and balanced in the management framework by the entrepreneurial figures present among the personnel of the enterprise, who are usually capable of expressing themselves with greater enthusiasm and trust.

The financial function constitutes a special case. It is often linked to the administrative sector, yet by its very nature it has greater similarity with the world of production. It is through the financial function that the financial means necessary for the management are procured and the allocation of any monetary deposits is carried out. This entails shouldering the responsibility of the related costs and profits, which may have a substantial influence on the trend of the profit and loss statement.

A clearer picture of these forms of behavior can be obtained by considering the nature of the information flows that arise during the management of an enterprise. At least two types of information systems can be identified, and although they necessarily have to be integrated, they travel along parallel lines but in opposite directions. This constitutes one of the causes of complexity.

The first system to be considered is the administration and management of information concerning facts and events that have already taken place (ex post), with a flow from the periphery toward the center. Here one finds centralization nodes with decentralized control, and the personnel working in this sector have mainly bureaucratic connotations.

The second system, which pertains to planning, is based on information concerning designs and entrepreneurial programs for the future (ex ante). Its action develops mainly from the center toward the periphery, with nodes that focus on operative distribution of information, but with centralized control. Management of this system is entrusted to typical entrepreneurial figures and develops by means of an information and communication system that has to allow control at the various critical threshold levels, identifying and defining possible points of dissension, above all in situations of an impending crisis.

Information flows are aided by the existence of some particularly useful and significant institutional organs. In particular, the "management committees" constitute an effective organizational tool that facilitates information transfer between the two systems. Management committees can also represent an efficient opportunity for coordination and control of the evolution of various aspects of management.

One important element to take into consideration is training, which should be regarded as a strategic function since it is crucial for the realization of a learning organization, i.e., for the creation of an enterprise that is constantly involved in learning, in line with the most significant strategies of corporate organizational evolution.

Complexity can thus be simplified by the use of network models, which exemplify the corporate structures that generate the resulting strategies. The model of a network enterprise and enterprise networks also facilitate the establishment of a concept of "normalcy," which is indispensable to untangle at least in part the labyrinths of complexity as well as the possible dangerous evolutions of complexity that may trigger veritable corporate crises.

Notes

1. Ronald H. Coase, "The Nature of the Firm," originally in *Economica* (November 1937), reprinted in *The Nature of the Firm: Influence*, in *The Nature of the Firm: Origins, Evolution and Development*, edited by Oliver E. Williamson and Sidney G. Winter, Oxford University Press, New York, 1993, pp. 386–405, and Ronald H. Coase, "The Problem of Social Cost," in *Journal of Law and Economics* 3 (October 1960), pp. 1–44.

2. Ronald H. Coase, *The Nature of the Firm*, cit., p. 390.
3. Ibidem, p. 391.
4. Ronald H. Coase, Ibidem, cit., p. 61.
5. *The Economic Nature of the Firm: A Reader*, edited by Louis Putterman and Randal S. Kroszner, Cambridge University Press, Cambridge, 1996.
6. Ronald H. Coase, *The Nature of the Firm*, cit., p. 386.
7. Oliver E. Williamson, *Market and Hierarchies: Analysis and Antitrust Implications*, The Free Press, New York, 1975; *The Economic Institution of Capitalism: Firms, Markets, Relational Contracting*, The Free Press, New York, 1985; *The Economic Institutions of Capitalism: Firms, Markets, Relations, Contracting*, The Free Press, New York and London, 1985; *The Mechanisms of Governance*, Oxford University Press, New York, 1996.
8. Oliver E. Williamson, Masahiko Aoki and Bo Gustafsson, *The Firm as a Nexus of Treaties*, Sage Publications, London, 1989.
9. Pietro Romano, Paolo Danese, *Supply Chain Management. La gestione dei processi di fornitura e di distribuzione* ["Managing Business Processes Across Supply Networks: The Role of Coordination Mechanisms," *Journal of Purchasing and Supply Management,* Digital Document, Elsevier 2004, http://www.sciencedirect.com/science?_ob=ArticleURL&_udi=B7579-4FBHW7M-2&_user=10&_rdoc=1&_fmt=&_orig=search&_sort=d&_docanchor=&view=c&_searchStrId=945307772&_rerunOrigin=google&_acct=C000050221&_version=1&_urlVersion=0&_userid=10&md5=bb2baa539edff152b59783e6f1cc83e6], McGraw-Hill, Milan, 2006, p. 79.
10. Oliver E. Williamson, *The Mechanisms of Governance*, cit., p. 212.
11. Ibidem, p. 7.
12. Michael E. Porter, cit., Edizioni Comunità, Milan, 1987.
13. G. Francesco Visconti, *La teoria dei costi di transazione e l'evoluzione del mercato del lavoro* ["Transaction Cost Theory and the Evolution of the Labor Market"], in *Bari Economica* n° 1, 2003, pp. 90–97.
14. Alfred D. Chandler, Jr., *Strategy and Structure: Chapters in the History of the Industrial Enterprise*, MIT Press, Cambridge, MA, 1962, p. 8.
15. Ibidem, p. 11.
16. Ibidem, p. 11.
17. Oliver E. Williamson, *The Mechanisms of Governance*, Oxford University Press, New York, 1996, ix.
18. Ilya Prigogine, *From Being to Becoming: Time and Complexity in the Physical Sciences*, W. H. Freeman & Co., San Francisco, 1980.
19. Federico Butera, *Sviluppo tecnologico e disoccupazione: trasformazione della società* [*Technological Development and Unemployment: Transformation of Society*], Accademia Nazionale dei Lincei, Rome, 1998, pp. 375–376.
20. Ibidem, p. 376.
21. Merritt Roe Smith and Leo Marx, editors, *Does Technology Drive History? The Dilemma of Technological Determinism*, MIT Press, Cambridge, MA and London, 1994.
22. Larry E. Greiner, *Evoluzioni e mutamenti nelle imprese che si espandono* ["Evolutions and Changes in Expanding Enterprises"], in *Quaderni di formazione Pirelli* [*Pirelli Training Papers*], n° 14, April 1975, p. 9.
23. G. F. Dioguardi, *Organizzazione come metodo di scienza e di impresa. Da Claude Bernard a Henri Fayol* [*Organization as a Method of Science and the Enterprise. From Claude Bernard to Henri Fayol*], Il Sole 24 Ore, Milan, 1988.
24. G. F. Dioguardi, *Crisi nella gestione dell'impresa* [*Crisis in Enterprise Management*], Dedalo, Bari, 1999.
25. J. de Rosnay, *Il macroscopio. Verso una visione globale* [*The Macroscope: A New World Scientific System*, Harper & Row Publishers, New York, Hagerstown, San Francisco, London, 1979], Introduction by Gianfranco Dioguardi, Dedalo, Bari, 1977, p. 68.
26. G. F. Dioguardi, *L'impresa come laboratorio. Esperienze, teorie, esperimenti* [*The Enterprise as a Laboratory: Experiences, Theories, Experiments*], Il Sole 24 Ore, Milan, 1986.

27. G. F. Dioguardi, *L'impresa nell'era del computer* [*The Enterprise in the Era of the Computer*], Il Sole 24 Ore, Milan, 1986.
28. R. D. Stacey, *Managing Chaos: Dynamic Business Strategies in an Unpredictable World*, Kogan Page Ltd., London, 1992, p. 180.
29. G. F. Dioguardi, *La natura dell'impresa fra organizzazione e cultura* [*The Nature of the Enterprise Poised Between Organization and Culture*], Laterza, Bari, 1996.
30. R. D. Stacey, cit., p. 81.
31. G. F. Dioguardi, *L'impresa nella società di Terzo millennio* [*The Enterprise in the Society of the Third Millennium*], Laterza, Bari, 1995.
32. R. D. Stacey, cit., pp. 182–183.
33. Furio Colombo, *La vita imperfetta* [*The Imperfect Life*], Rai-Eri-Rizzoli, Milan, 1999, p. 91.

Chapter 3
Innovation and Service Provision in the Network Enterprise

Among the phenomena that characterize the network enterprise in the high-tech world, one important aspect concerns its marked propensity toward tertiarization. In fact one of the most typical indicators of the economic development of a country consists precisely in the development of the "service" sector, i.e., the so-called process of tertiarization whereby the sector of services and commerce progressively tends to prevail over that of the products produced by agriculture and industry.

The concept of "service" is not easy to define, although its meaning is well known in the common use of the term. The *Grande Dizionario della Lingua Italiana* [Complete Dictionary of the Italian Language] provides a number of definitions, including some with a religious meaning. It also offers an explanation that makes reference to *public services*, stating that these represent "Any performance of service provided by the State, by a local administration or by an enterprise that holds a concession, which is designed to satisfy a need felt by the community as a whole, either of a material or cultural and social nature [. . .]."[1] And with regard specifically to the economic meaning of *service*, this is defined as "Any performance of a service that is designed to satisfy an individual or collective need and which can be assessed in economic terms and in terms of being bought and sold [. . .]."[2] The dictionary goes on to specify that the *Service Sector* (also known simply as "*Services*") is composed of the "set of economic activities the object of which is not the production of new goods, but rather the execution of work or tasks in the field of commerce, transportation, education, the state administration, etc.; tertiary sector [. . .]."[3] On this subject the *Enciclopedia Europea* [European Encyclopedia] specifies that "[. . .] In addition to the traditional sectors of commerce and transportation, and credit, other activities are coming to the fore and are playing an increasingly important role: such activities include data elaboration and information processing (IT), data transfer (telecommunications), marketing and customer care, leasing, scientific and technological research (research and development). These are activities that are not restricted to enhancing and streamlining the process of product marketing: rather, they are becoming increasingly essential and crucial for the administration and management of enterprises themselves, as well as for the future prospects of enterprises (suffice it to mention the role of scientific research, which constitutes one of the most momentous drivers of economic and social progress."[4]

G. Dioguardi, *Network Enterprises*, Innovation, Technology, and Knowledge Management 104, DOI 10.1007/978-1-4419-1333-3_3, © Gianfranco Dioguardi, 2010

Thus the definitions tend to shift attention away from the macroeconomic area that refers to the aggregate units of investigation and toward management of the individual enterprise. The new perspective is set within the overarching context of the sweeping organizational transformations that have affected enterprises in recent decades, effectively tertiarizing a substantial proportion of the activities of an enterprise. This phenomenon has undergone rapid development with the rise of the IT revolution and the advent and spread of the world of high-tech, in particular the technology linked to electronic data processing (*EDP*) and information and communications technology (*ICT*). Within the enterprise, the electronic data processing (EDP) service and the related information services constitute typical elements of a high degree of tertiarization, one of the effects of which has been to introduce rather radical changes in management of the "administration function." But the other corporate functions are likewise undergoing pronounced transformations that are moving in the same direction.

For example, one interesting ongoing change concerns the "purchasing function." This function is being affected by a twofold transformation, in that it is becoming both strategic and tertiarized at one and the same time. Thus it is being called on to exert control over what is now a vast range of contracts linked to the presence of the macroenterprise, the latter being composed of the complex network of sub-suppliers and services that the reference enterprise buys in from third parties upstream from actual in-house production. In this perspective, Romano and Danese propose a theoretical account of supply chain management, clarifying its meaning: "it is therefore necessary for the supply chain manager to carry out an analysis of all the products bought in by the enterprise, so that it becomes possible to select, for each product, the ideal type of relation with the given supplier."[5] This manner of proceeding is facilitated by the fact that the production function is experiencing tertiarization as a result of the widespread use of robotics and IT applied directly to production processes.

As far as the sales function is concerned, this is often carried out through communication networks such as the Internet or the Intranet. More generally, sales data management and information on customer preferences, to be used as a guide to changing demand, are managed by increasingly automated services.

The marketing function, which is one of the most typical tertiary services, has acquired increasing prominence in the corporate context over recent decades, to the point of taking on a priority role. Today, the marketing function serves not merely as a means of promoting sales but above all as a route-finder which coordinates the positions of the various customers. It thus becomes possible to analyze the various different requirements that characterize the demand, thereby setting the stage for the qualitative and quantitative characteristics of products the enterprise will subsequently produce.

The personnel function is managed through the corresponding databases, allowing all the processes concerning the workers' activities to be automated with the aid of purpose-designed services.

In addition, management of the function in charge of corporate organization is likewise becoming tertiarized, inasmuch as it is strongly oriented toward preliminary planning processes, which are becoming increasingly important with respect to the moment of actual execution.

Particular importance attaches to one of the most typical tertiary functions, namely that of finance, which has acquired momentous importance in recent times, becoming almost more important than the production and sales functions themselves, given the significance and ever-increasing development of the capital market. Management control is also of mounting operative significance due to the possibility of making use of services typical of the advanced tertiary sector. Above all, this sector has seen the rise of new systems of appreciation of performances and quality control, leading to highly innovative enterprises within the economic scenario.

Among the other activities that are strongly swinging the enterprise toward tertiarization is the research and development function, which is now considered indispensable to assure corporate survival on unstable and turbulent markets. Today's volatile markets make it vital for enterprises to respond promptly to variations in demand, with the result of sparking constant innovation, both in process and in product. This underlines the present-day relevance of the Schumpeterian definition of the enterprise as a tool for introducing innovation into the economic system.

Indeed, today more than ever before, what is highlighted is the propensity of entrepreneurial innovation to become process innovation, in the sense that the very concept of product is undergoing a modification toward a form of service provision expressed by its utilization. This awareness represents a new frontier for current innovative research.

It is of vital importance that the settings which form the wellspring of innovative inspiration and the related venues of advanced training should become "poles of excellence," represented fundamentally by university institutions located in a number of different areas. By acting as a driving force within their community, such institutions can provide support to enterprises – especially small- and medium-sized businesses – and thus help to devise survival strategies that will enable such enterprises to grow and expand into global markets.

This revisitation of the concept of "innovation" calls to mind once again the definition that was given by Schumpeter in his analysis of the enterprise. According to Claudio Napoleoni in his celebrated *Dizionario di Economia Politica* "What is known as an enterprise is in effect the introduction of an *innovation* into the economic system. Innovation [...] is defined by Schumpeter as a 'new combination' of production factors: that is to say, innovation takes place every time the production process undergoes a real qualitative modification. Schumpeter further suggests that the concept of innovation includes the following cases: (1) Introducing a new commodity [...]; (2) Introducing a new production method, [...], which [...] may consist, for example, in a new way of dealing with a product commercially; (3) The opening up of a new market [...]; (4) Securing a source of the supply of raw materials and semifinished products independently of whether such a source was already in existence or had to be created *ex novo*; (5) establishing new types of organization of a certain industry [...]." In short, "A person who introduces an innovation is, at the moment it is introduced and by the sheer fact of introducing it, termed an 'entrepreneur.' "[6]

It becomes clear, then, that by its very nature an enterprise must be supported by innovation that is sustained over time and constantly replenished both in its quantitative and qualitative dimensions. This is of the utmost importance today, in a context of turbulent scenarios that trigger rapid changes affecting all the rules of the economic game and of ever fluctuating trends on the exchange markets. Only by a constant response characterized by the distinctive feature of innovation can an enterprise cope with this dimension of change.

It is also worth noting that in the present state of affairs, the search for innovation in the Schumpeterian sense appears to be facilitated by the fact that products have in general ceased to exclusively assume a configuration that could be described, as it were, of hardware. Rather, the products supplied by the enterprise system, in particular high-tech products, are increasingly being transformed into what could be described as process products. They can therefore be likened to veritable services that are made possible by their provision of utilization, since they are almost always characterized by an immaterial component definable as knowledge: a form of knowledge linked to the technological complications of the manufacts and thus to their intrinsic complexity. In other words, it is knowledge that is necessary to facilitate interaction with the user and consequently to allow the best utilization of such products. Furthermore, it is also knowledge that is needed for optimal conservation of the products through innovative maintenance processes.

These aspects can be metaphorically considered as the necessitated consequence of the contribution made by "knowledge workers" to management processes, in the sense that such workers, who operate in modern high-tech "network enterprises," produce service products which, inescapably, are at the same time definable as knowledge products. Such products are endowed with an intensity of knowledge that allows them to configure themselves by means of physical hardware which, increasingly, is accompanied by cognitive software indispensable for actual utilization of the products in question.[7]

In this perspective, product innovation today often translates into veritable processes capable of extending both the production process and also the marketing process. The latter is linked to sales and real exchanges, which are themselves becoming more and more complex on account of the sophisticated communication tools used in these sectors. In effect, within the current framework the entire Schumpeterian cycle of innovation is strongly linked to ICT, i.e., to advanced technology in the field of information processing and communication – or, more generally speaking, to knowledge in the broad sense.

Casting one's eye back over the history of the enterprise, it is interesting to note that in its early stages the enterprise was a closed entity, with a specific propensity to carry out the entire production in-house according to the typical *make* approach. Moreover, its activity took place in a context almost exclusively linked to material production.[8] In such a situation, services were virtually absent, or at the very least they were of only minor significance. And production, based as it was on very large volumes, was – as the term itself suggests – mass production, which tended to award priority to quantity as compared to diversification of quality. This mass production was achieved in the factory by breaking actions down into a multiplicity of elementary tasks, which could be performed by workers whose training was rudimentary. In fact, it is no exaggeration to state that in such work places the concept of knowledge and culture as a service for the individual was practically absent, its only traces appearing in the form of corporate culture, which effectively meant simply the historical background of the enterprise.

But the old rules of the game have been dramatically changed by the introduction and increasing spread of automated machines on the assembly line – in particular computers and also robots – as well as by the advent of distributed IT and advanced technology in the various corporate functions. A major thrust in this direction has also come from communications technology, which has enabled the markets to go global. Production organizations have opened up to external sources and have begun to operate mainly with the *buy* approach, namely buying in from the outside many of the parts and services needed for production. Such transformations have proved fundamental in rendering productions increasingly variable, thereby giving rise to flexible or "lean" enterprises.

One of the interesting offshoots of this situation is that individual culture has once more begun to play a fundamental role. No longer is culture conceived merely as the history of the enterprise (corporate culture): rather, culture is now felt to express the history of the individual and individual behavior, in a context capable of rediscovering quality but also the values of ethics. The ethical approach starts out from consideration of the individual worker and can evolve into a veritable form of corporate ethics.

Overall, then, the enterprise has increasingly undergone a transformation in the direction of tertiarization, and within this framework the world of service provision has begun to play an ever more significant role.

Turning now to the network concept, this is expressed above all in terms of the great organizational flexibility and variety of products that can be achieved by implementing as intensively as possible the policy of buying from the outside a large number of parts and services required for the production cycle. These, in turn, give rise to a network of third enterprises, external to the reference enterprise. Taken as a whole, these external enterprises contribute in an integrated manner to the production cycle and thus to the organization of the mother enterprise, thereby forming the macroenterprise. As a first step in building up the macroenterprise, the mother enterprise generally sets up a series of contacts with external enterprises, followed by actual contracts. Such procedures are the typical elements of pronounced tertiarization of relations upstream from production itself. The contracts have to be monitored and enforced by an appropriate control system, which deals both with quality control and respect of the clauses of the contract, especially delivery deadlines and agreed manner of delivery, both of which must be rigorously respected.

Basically, then, tertiarization transforms the enterprise into a sort of "laboratory" of continuously innovative experimentation. This results in a *network* of relations which, taken together, constitute a many-stranded "web" with a reference subject represented by the modern manager, who is capable of expressing a specific action of governance over these relations.

Flexibility, which is typical of the organizational structure of the enterprise, prompts a substantial change in the operative strategies adopted today in facing the market and demand, especially the potential sources of demand that are likely to become customers of the enterprise. In this perspective, the customer becomes a protagonist of the production process, and flexibility allows the enterprise to adapt continuously to the changing requirements of a demand which is itself also constantly changing, in the wake of an ever-increasing need for services that can enrich the performance of the products.

As a consequence, a growing degree of complexity is embodied in the actual utilization of innovative products. This prompts the enterprise to supply nothing short of genuine information and training processes as an intrinsic accompaniment of a product, to enhance product utility, as well as to ensure optimal maintenance for better preservation of the products themselves. Furthermore, each product becomes the object of intense marketing campaigns, which range from a simple advertisement or commercial spot to sophisticated forms of propaganda that seek to involve the consumer in an increasingly global manner. Thus the product is transformed into something strongly resembling a process, where the tertiarized service component may actually prevail as compared to the product in itself.

It is worth recalling the way the great American scholar of business history Alfred D. Chandler interprets these phenomena in his celebrated 1977 book *The Visible Hand: The Managerial Revolution in American Business.* Chandler considers the modern enterprise as having replaced traditional market mechanisms, which, harking back to the theory put forward by Adam Smith, were guided in the past by the "invisible hand" of the egoism of individual players as they engaged in coordination of economic activities and allocation of resources.[9] Chandler argues that in many sectors of the economy the "visible hand" of the managers has replaced the above mechanisms and has crucially affected the characteristics of demand and consumer choice.

These considerations suggest that the phenomenon of tertiarization is becoming widespread not only within the production enterprise but also upstream and downstream of its production processes. This is becoming particularly evident downstream, with a pronounced transformation in relations with customers, who can more and more be seen as the recipients of services rather than merely the recipients or purchasers of products.

In effect, what originally started out as something that characterized IT products, which were composed of a material element – the hardware – that could not be used unless it was supplied with a purpose-designed program – the software, has now spread through almost the whole range of products. The real priority, in today's context, is the use of products as a service provided to satisfy the needs for which they were designed.

This new scenario, however, is not without its problems. In the tertiarized enterprise a curious phenomenon is emerging which needs to be carefully monitored: namely, a sort of "neo-Taylorism," but this time of an intellectual nature. In all of these organizations, operators now work in a close relation with IT machines, in particular computers. In fact, one can go so far as to say that man and machine now constitute a remarkable indissoluble couple, composed of a human individual and a computer, with many of the operator functions being delegated to the computer. Thus the activity is carried out through what was defined earlier as a "technological assignment of responsibility," ranging over a growing spectrum of functions. This results in even more intense tertiarization of the overall situation, and the individual's concentration on the tasks and controlling actions to be performed is attenuated.

This question of the dulling of the senses when engaging in the use – or indeed the abuse – of technology is one of the most worrying aspects of the tertiarization of the organizational world. It needs to be countered in various ways, and one of the most effective forms of counteraction is by injecting abundant and carefully targeted doses of culture that will reinforce and reassert the operator's personality and human nature, and thus in a sense "reset" the operator's mindset. The operator will thus be enabled to continue to dominate the so-called smart machines without, however, being dominated and exploited by the latter. This implies that within the framework of the network enterprise, all individuals must learn to operate as if they were "self-entrepreneurs," seeking to revive the old enterprise spirit. By the same token, there is a need to counteract the frequently detrimental effect that arises if a state of affairs that could be likened to a "Baroque court" prevails. The "Baroque court" effect may come about if the individual operator's egoistic frame of mind predominates over the community spirit, discouraging any sense of performing a service in the interests of a higher cause shared by all those involved. Accordingly, what is called for is an awareness of individual ethics, which must contribute to shaping a "social responsibility of the enterprise" as a further manifestation of an important service provided by the institution within the social context.

The entrepreneurial framework as outlined above highlights the necessity to introduce new forms of management control, more clearly oriented toward a true culture of service. It has already been pointed out that in its transactional aspects the network enterprise is composed of a series of customers, suppliers, users, and utilizers both within and outside of its network of transactions, and that such aspects are generally regulated by contracts – especially with respect to the external world – and also by agreements or procedures within the management processes.

The resulting picture is one of a far-reaching mass of services with a maze of aspects to be controlled, such as performance efficiency, cost-effectiveness, or quality of the results. This in turn leads to the concept of new "rights" of the customers, but also of the suppliers – for example, to be drawn up in effective and exhaustive documents that clearly testify to the quality of the service provided – in order to be able to concretely assess the actual performance achieved. A further offshoot of this new state of affairs is an evaluation of "duties" in terms of respecting the services that are to be provided (suppliers) or purchased (customers).

An explicit overview of the information stating what type of performance is actually achievable will help to optimize clarity of relations between the parties. If the overview is mapped out while the execution is under way, it will be able to improve achievement of the final objectives. In this sense, there will be greater potential for control over a planning process, with the possibility of splitting up the complex events into a chain of specific objectives, to be accomplished while the actions themselves can still be subjected to improvements. In addition, explicit forms of guarantees will also be introduced, giving the players involved in the transactions the reciprocal power to bear witness to the strengths and weaknesses of the products and services.

The servicing of existing contracts must be managed in such a way as to optimize the results by means of continuous monitoring, so that the quality of performances, productivity, and management of customer–supplier relations can benefit from constant improvement. To ensure that proper steps are taken in this direction, enterprises should introduce measures that will build up a system of clear certification of performance, so that the parties involved can understand, check, and implement service quality. Great effort should be directed toward operating with maximum efficiency in terms of cost-effectiveness and quality. Analysis of quality and continuous control to ensure that contractual clauses are respected should therefore lead to a service of total data integration, making it possible at all times to obtain a general overview and thus to optimize corporate management.

It is useful, at this point, to take a fresh look at control of entrepreneurial services, examining the possibility of expressing explicit and unbiased evaluations of the various members of staff in charge of the different operations. A system could be introduced whereby merit would be rewarded by continuity of the currently existing relations.

But in order to appreciate more fully the innovative connotations of the definition of network enterprise, it is helpful to sketch the evolution of the various organizational theories over time and to report briefly on the types of experiences that have been documented.

In a general sense, organization theories spring from economic thought and take the form of events occurring in specific revolutions. Such events are then succeeded by evolutionary states capable of normalizing the stages that have been reached. These situations usually represent the modifications that have effectively been given concrete form within the framework of entrepreneurial management. According to Azzone and Bertele, "The organizational set-up – here this term includes the structuring of the organization *strictu senso,* the mechanism involved in its functioning and the corporate architecture of the enterprise – is naturally subject to variations over time: these may be fairly minimal in certain periods, or exhibit pronounced solutions of continuity in others."[10]

Adam Smith was the forerunner of the thought later given fuller expression by Taylor, who interpreted the so-called scientific organization of labor in the Fordist factory, while Fayol, a disciple of the French enlightenment, would subsequently build up the theories concerning the administrative governance of the enterprise. Toward the end of the 19th century the thought of these two great masters was to inspire a vast array of doctrines on the nature of the enterprise, often complementary to one another.

In a later age, the advent of *Information and Communication Technology* (ICT) would signal another revolution in organizational theories, prefigured as early as Ronald Coase and then developed in greater depth by Williamson and many others. The enterprise was now interpreted as a hierarchy capable of governing transactions, and it is in this context that the term "transactional enterprise" gained credence. The model became manifested in the form of a network of contracts, opening up new roles for corporate culture and behavioral ethics, for both the individual and the entrepreneurial unit. The expression "short enterprise" began to be used, referring to the fact that vertical interconnections in the organization chart declined drastically on account of the introduction of a flat structure. Eventually this form of setup gave rise to the "network enterprise" as the expression of an organizational unit that has been streamlined and rendered flexible in all its operative activities.

In order to gain greater insight into this complex set of developments, it is worth dwelling on the various historical stages that lie behind its present-day outcome.

Notes

1. *Grande Dizionario della Lingua Italiana* [Complete Dictionary of the Italian Language], S. Battaglia, editor, Utet, Turin, 1994, vol. XVII, p. 781 ff.
2. Ibidem.
3. Ibidem.
4. *Enciclopedia Europea* [*European Encyclopedia*], Garzanti, Milan 1980, vol. X, p. 433.
5. Pietro Romano, Paolo Danese, *Supply Chain Mangement. La gestione dei processi di fornitura e di distribuzione* ["Managing Business Processes Across Supply Networks: The Role of Coordination Mechanisms," *Journal of Purchasing and Supply Management*, digital document, Elsevier 2004, http://www.sciencedirect.com/science?_ob=ArticleURL&_udi=B7579-4FBHW7M-2&_user=10&_rdoc=1&_fmt=&_orig=search&_sort=d&_docanchor= &view=c&_searchStrId=945307772&_rerunOrigin=google&_acct=C000050221&_version=1&_urlVersion=0&_userid=10&md5=bb2baa539edff152b59783e6f1cc83e6], McGraw-Hill, Milan, 2006, p. 113.
6. Claudio Napoleoni, *Dizionario di Economia Politica [Dictionary of Political Economy]*, Edizioni di Comunità, Milan, 1956, p. 683.
7. Federico Butera, Enrico Donati, Ruggero Cesaria, *I Lavoratori della Conoscenza - Quadri, Middle Manager e alte professionalità tra professione e organizzazione* [*Knowledge Workers: Executives, Middle Managers and Professional Expertise in the Context of Profession and Organization*], Franco Angeli, Milan, 1997.
8. Federico Butera, *Il castello e la rete* [*The Castle and the Network*], Franco Angeli, Milano, 1990, and also Federico Butera, *Il campanile e la rete: L'electronic business e le piccole e medie imprese in Italia [The Belltower and the Network: Electronic Business and Small and Medium-Sized Enterprises in Italy]*, Franco Angeli, Milan, 2001.

9. Alfred D. Chandler, *The Visible Hand: The Managerial Revolution in American Business*, The Belknap Press, Harvard, Cambridge, MA, 1977.
10. Giovanni Azzone, Umberto Bertelè, *L'impresa: sistemi di governo, valutazione e controllo* [*The Enterprise: Systems of Governance, Evaluation and Control*], Second edition, Etas, Milan, 2003, p. 34.

Chapter 4
Outline of the History of Organization Doctrines

The course of history charts a path that is by no means uniform and unfaltering: rather, it proceeds by fits and starts which, through the succession of daily events, assume varying degrees of significance and not infrequently exert a decisive influence over the scenario and the future.

Awareness of such events and their description determines what we define as history, while their critical analysis underlies the establishment of philosophy, which addresses the problem of knowledge. Philosophy extends the scope of knowledge beyond the bounds of metaphysics, so that its interpretation embraces the sphere of aesthetics, ethics, law, economics, organization, sociology, and politics.

Philosophy thus offers itself as the most significant and most wide-ranging understanding of human wisdom. From this source there have sprung a number of offshoots, which in turn have burgeoned into the individual sciences that can be perceived in embryonic form in the thought of the great philosophers, from the Greek masters to the protagonists of modern inquiry.

Philosophy prompted investigation of the physical world, namely, that mode of investigation which rose to the dignity of science in the early 17th century when Francis Bacon (1561–1626) introduced the experimental method and Galileo Galilei (1564–1642) developed the language of mathematics, becoming the first scientist capable of attributing quantitative dimensions to phenomena. Science is here taken in the sense of a "body of rigorously controlled and systematically ordered forms of knowledge by virtue of which it becomes possible to reach objective truths concerning a given order of phenomena or concepts."[1]

Meanwhile William Petty (1623–1687) had laid the bases of economics as a science, by discussing the issue of political arithmetic, or, as Charles Davenant (1656–1714) would later put it, "the art of reasoning through numbers on matters concerning government [...], undoubtedly a very ancient art."[2] It was Petty who first introduced rules and a systematic method for quantitative analysis of phenomena. This marked the primordial stages of economy, which would take on an independent dimension in the 18th century when, among other developments, chairs of this discipline began to be established in universities. In this regard, the figure of Antonio Genovesi represents an important landmark, being the first to hold public lessons of economy and commerce in Naples in 1754.

G. Dioguardi, *Network Enterprises*, Innovation, Technology, and Knowledge Management 104, DOI 10.1007/978-1-4419-1333-3_4, © Gianfranco Dioguardi, 2010

Organization theory subsequently arose from the economic contexts, but it was not until the 20th century that it achieved a recognized independent status, when quantitative methods and techniques were invoked for study of this field as well. Links, albeit often rather tenuous, continued to be maintained both with economy and philosophy, from which the theory had drawn its origin, revealing glimpses of "the hidden connections," to quote Fritjof Capra.[3]

Thus the grandiose events of the world – whether they be happenings or theoretical intuitions – could be defined as "events" capable of producing, or helping to produce, veritable "revolutions," which in the fullness of time are followed by the evolution of the ideas and facts that have sprung from such revolutions. As the turmoil of revolution quietens, the new conditions become more clearly defined and multiply in expanding waves of the development of civilization. This transformative diffusion was very aptly described by an Italian historian – Franco Venturi – when he commented that "chancing on the birth of a new way of feeling and thinking, capturing the moment when a new element becomes part of a civilization and begins to modify and transform it is perhaps the most passionately exciting thing a historian can hope to achieve."[4]

Revolutions that are manifested in the form of new theories or noticeable changes in daily routines are present in the history of the various individual disciplines that have arisen as branches of human knowledge. It is therefore hardly surprising that they have left their mark on organization theories as well. Consequently, the statement by Thomas S. Kuhn in his celebrated book *The Structure of Scientific Revolutions*, which was intended by its author to refer specifically to science, can be taken as holding in an absolutely general manner.

Kuhn's argument is that when scientific theories are considered in an evolutionary perspective, they can be seen to be characterized by the presence of revolutions conceived as changes in the conception of the world; moreover, such changes are capable of generating paradigms that act as the overarching frame for "normal science," i.e., for " research firmly based upon one or more past scientific achievements, achievements that some particular scientific community acknowledges for a time as supplying the foundation for its further practice."[5] Thus normal science supplies theories and models that belong to specific "paradigms," the latter being understood as "the entire constellation of beliefs, values, techniques and so forth shared by the members of a given community."[6] Moreover, Kuhn points out that "Rules [. . .] derive from paradigms, but paradigms can guide research even in the absence of rules."[7]

This leads to a situation featuring cycles represented by the succession of theoretical discoveries and inventions, which in turn give rise to theories capable of prefiguring further innovations. This phenomenon – as has been noted – takes on very general connotations inasmuch as it is applicable to the development of any expression of human cognition or any branch of human knowledge whatsoever. Effectively, whenever a pathbreaking innovation makes itself felt, it leads to a paradigmatic change that is nothing short of a genuine revolution. Such changes occur at various moments in history, partly shaping the development of history itself, and in any case influencing the course of history. The effects may make themselves felt in the "short term," the "medium term," or the "long term," as depicted by Fernand Braudel, whose account was further developed by George Duby in the following illustrative explanation: "[. . .] duration, event, conjuncture and structure (the events, on the surface, like foam, overlying the oscillations of the conjuncture, and, as a scaffold supporting the entire edifice, structures, being imperceptibly drawn and tugged by slower movements, the latter two terms of this ternary figure, conjuncture and structure, having been borrowed – let us take note – from the language of economics)."[8]

Revolutions need not, however – and indeed only rarely do – burst onto the stage as a sudden full-blown upheaval: they may appear on the scene as creeping change, a gradual accumulation of situations or theoretical hypotheses that are almost imperceptible at the moment of occurrence, but which gradually build up over time until a threshold limit is exceeded. Only at this point is what René Thom called "catastrophe point" reached, with its potential for macroscopic disruption of preexisting situations.[9]

All this did indeed take place in the sciences and in the disciplines that deal with the economy in general, and thus also in organization theory. What is more, one should not be misled into thinking that epoch-making changes occurred only at the dawn of scientific development, when scientific inquiry was driven above all by the art of waging war which by its very nature revolutionizes daily life. All the branches and disciplines of life have been characterized by revolutions and evolutions which have in most cases developed over fairly prolonged time periods. Interestingly, though, one significant feature of the late 20th century – the century denominated by Eric J. Hobsbawm's *The Short Century* – was a drastic reduction in the length of such incubation periods, with changes succeeding one another at an amazingly rapid pace.[10]

In history, starting from medieval economy, which was characterized by scarcity and drew its inspiration from religious teachings, after a prolonged period of time the slow but constant modifications of the medieval vision of life eventually led to the emergence of the extraordinary revolution that was the Renaissance. The protagonist of the productive world was at that time the craftsman, who produced a myriad of different objects in his workshop. These were usually high-quality objects, with the quality directly controlled by the craftsman himself. It was in this world and this era that the figure of the merchant came to the fore, a figure who can be seen retrospectively as having been the first to interpret the role of the bourgeois capitalist.

According to Piero Ottone, the organizational philosophy of the "mercantilists" – the merchant economists – "consisted in selling a large quantity, selling at a higher price than was paid out for the original purchase, and selling rapidly."[11] This provided the rationale for extensive foreign trade. A celebrated case is that of the commercial network Francesco di Marco Datini (1135–1410) – the Merchant from Prato – succeeded in setting up throughout what was then the whole of the known world. It could almost be said that he was a forerunner of the modern *holding*, with its various branches activated through the participation of local experts.

Thus the mercantilist economy was founded on the figure of the merchant as an interpreter of the trade in arts and crafts, a form of production whose organization took the form of the workshop-factory. The subsequent revolutions in practice and theory would profoundly change these figures and their very places of work, introducing the concepts of the entrepreneur and manufacturing which would subsequently evolve into modern industry.

This led to a gradual change from purely commercial capitalism to industrial capitalism: "often, it was the merchants themselves who engaged in a form of industrial activity: first they would begin by distributing the raw materials to independent workers, who carried out the processing on their behalf within their own homes, then later the merchants found it more advantageous to gather the workers together in factories."[12] Indeed, the project of the industrial factory delineated in this description also heralded the subsequent step that would supersede it through the spread of the concept later destined to prefigure the "cottage industry," namely the idea of working from home, distant from the industrial factory.

Thus there emerged the new figure of the "entrepreneur," and it was Richard Cantillon (1680–1734) who would for the first time introduce the term *entrepreneur* into the language of economics. This was defined in his *Essai sur la nature du commerce en général*, published posthumously in France in 1755, in a very broad context that allowed him to include even the classes of beggars and robbers in this category.

The historical importance of the author and the clarity of his thought suggest a revisitation of the passages where Cantillon dwells on the entrepreneur. They appear in Chapter XIII entitled "The circulation and exchange of goods and merchandise as well as their production are carried on in Europe through by Undertakers, and at a risk." Here are some examples: "The Farmer is an undertaker who promises to pay the Land-owner, for his Farm or Land, a fixed sum of money [. . .] without assurance of the profit he will derive from this enterprise [. . .] [H]e carries [his produce] to market, or sells in the Market of the nearest Town, or perhaps there are a few individuals who set up as Carriers themselves. These bind themselves to pay the Farmer a fixed price for his produce, that of the market price of the day, to get in the City an uncertain price in the city, which should however defray the cost of carriage and leave them a profit. But the daily variation in the price of produce in the City, though not considerable, makes their profit uncertain.

The Undertaker or Merchant who carries the product of the Country to the City cannot stay there to sell them retail as they are consumed. . . [. . .] For this reason many people set up in a City as Merchants or Undertakers, to buy the country produce from those who bring it or to order it to be brought on their account. They pay a certain price following that of the place where they purchase it, to resell wholesale or retail at uncertain price [. . .]. These Undertakers can never know how great will be the demand in their City, nor how long their customers will buy of them, since their rivals will try all sorts of means to attract customers from them. All this causes so much uncertainty among these Undertakers that every day one sees some of them become bankrupt [. . .] Shopkeepers and retailers of every kind are Undertakers who buy at a certain price and sell in the Shops or the Markets at an uncertain price.

[...] All these Undertakers become consumers and customers one in regard to the other. [...] They proportion themselves in a State to the Customers or consumption. [...] [E]xcept the Prince and the Proprietors of Land, all the Inhabitants of a State are dependent; [...] they can be divided into two classes, Undertakers and Hired people; and the others on wages fixed so long as they receive them though their functions and ranks may be very unequal.[...] All the rest are Undertakers, whether they set up with a capital to conduct their enterprise, or are Undertakers of their own labour without capital, and they may be regarded as living at uncertainty; the Beggars and even the Robbers are Undertakers of this class."[13]

The entrepreneur, essentially engaged in commerce, was thus a person who made purchases at prices – i.e., at costs – that were defined and specified in order to sell at prices – i.e., at profits – that were uncertain. Consequently, one of the essential characteristics of the entrepreneur was that of taking on an entrepreneurial risk. It was a risk that plunged enterprises into a state of perennial uncertainty, yet it was also capable of generating profit, as theorized for instance by Frank H. Knight (1855–1972) in an important work published in 1957, *Risk, Uncertainty and Profit*, which mentions, among other things, skill and luck as being among the peculiar endowments of the entrepreneur.[14]

The Italian writer Gianmaria Ortes (1713–1790), in his essay *Dell'economia nazionale* (1776), formulated a definition of the "negotiation entrepreneur," in which he attributed to this figure a flair for imagination as a complement to the characteristic of the intellect typical of operative practice. Ortes thus laid the basis for the present-day distinction between entrepreneurs and managers. He also discusses the difference between the owner and the capitalist, which Marx would later take up again, casting the discussion in terms of officials of capital rather than entrepreneurs.

In contrast to Marxian theories, the heretical writings of Aleksandr Alexandrovïc Bogdanov (1873–1928) featured a broad-ranging treatise composed between 1913 and 1917 that dwelt on the question of tektology considered as the general science of organization, in which he suggested that knowing how to organize the means of production was the most important aspect of the ownership of such means.[15] In fact Jean-Baptiste Say (1767–1832), in particular in his *Treatise on Political Economy*, dating from 1803, had already outlined the figure of the "entrepreneur of industry" as the coordinator of production factors.

Thus the growing body of economic knowledge began to delineate the figures which, in later times, were to become the protagonists of industrial organization. The associated theories that would be put forward to offer a systematic account of this form of organization also gradually began to take shape.

Among the various conceptions that foreshadowed theories on organizational processes, it is also worth recalling the "Physiocratic" approach. With their motto of *laissez-faire, laissez-passer*, the Physiocrats endorsed "for the first time the independence of economic life from political establishments."[16] In their perspective, "the value of an industrial product is only the sum of the pre-existing values, that have been blended together in the product [...]; the only true creation of wealth is that of agriculture, it is that of the land."[17] The leading figure among the Physiocrats, François Quesnay (1694–1774) – the doctor who attended to Madame

Pompadour – composed his celebrated Tableau Économique (1758), giving an analytical representation of the economic system based on the division of society into three social classes, which Roncaglia and Sylos Labini thus summarize in a well-known paper: "propertied classes, (nobility and the clergy), productive classes (entrepreneurs and farmers), sterile classes (craftsmen, manufacture workers). Between these classes a linkage had arisen in the form of an exchange circuit, leading at one and the same time to income distribution among the different classes and to the condition whereby the production process was repeated in a regular fashion. Economic activity was thus conceived as a 'circular flow' of production and exchange, a conception that characterized the entire field of classical political economy up to Marx and which [. . .] has been reproposed in recent times by Pietro Sraffa."[18]

Also prefigured in Quesnay's work were the sectorial interdependence tables that Wassily Leontief would later apply to economic systems. These have been utilized in some of the very recent network enterprise models as well, where they serve to interpret internal transactions relating to corporate production processes.

In effect, at least initially, economy also contained within itself, almost implicitly and thus with "hidden connections," observations of an organizational character concerning the protagonists of the production process. This is particularly true of the thought of Adam Smith (1723–1790), whose work "deserves to be remembered first and foremost as a highly rational contribution to economic theory."[19]

Smith, a great systematizer of preexisting economic concepts, can effectively be considered as the founder of classical political economy.[20] The rise to prominence of this scholar marks a veritable revolution in economic knowledge. At first this took the form of a radically innovative evolution of established concepts, which then rippled through the various schools of thought that had little by little arisen in the wake of the new ideas. According to the subdivision proposed by William J. Barber, "In the history of economic ideas, as it has taken shape over the course of history, four major analytical traditions – the classical, Marxian, neoclassical and Keynesian – stand out. Each was organized around a different set of questions."[21] Each of these schools can be associated with a "revolution" that changed the paradigm within which economic thought understood as a form of scientific training was being shaped.

In the organizational framework, Adam Smith – addressing the issue of organization in his famous pin factory – laid the bases for organization science. It is of notable interest to reread this memorable page of economic history and organizations, taken from the chapter signally entitled "The division of labour": "The greatest improvement in the productive powers of labour, and the greater part of the skill, dexterity, and judgment with which it is anywhere directed, or applied, seem to have been the effects of the division of labour. The effects of the division of labour, in the general business of society, will be more easily understood by considering in what manner it operates in some particular manufactures. [. . .] To take an example, therefore, from a very trifling manufacture; but one in which the division of labour has been very often taken notice of, the trade of the pin-maker; a workman not educated to this business (which the division of labour has rendered a distinct trade), nor acquainted with the use of the machinery employed in it (to the invention of which the same division of labour has probably given occasion), could scarce, perhaps, with his utmost industry, make one pin in a day, and certainly could not make twenty. But in the way in which this business is now carried on, not only the whole work is a peculiar trade, but it is divided into a number of branches, of which the greater part are likewise peculiar trades. One man draws out the wire, another straights it, a third cuts it, a fourth points it, a fifth grinds it at the top for receiving, the head; to make the head requires two or three distinct operations; to put it on is a peculiar business, to whiten the pins is another; it is even a trade by itself to put them into the paper; and the important business of making a pin is, in this manner, divided into about eighteen distinct operations, which, in some manufactories, are all performed by distinct hands, though in others the same man will sometimes perform two or three of them. [...]The division of labour, however, so far as it can be introduced, occasions, in every art, a proportionable increase of the productive powers of labour."[22]

It was, however, with the advent of the industrial revolution that organizational problems became dramatic. By virtue of its technological, economic, and social components, industry was now able to concentrate enormous quantities of energy, machines, and manpower within well-delimited spaces. And with the rise of the industrial system the most keenly felt and widely discussed problems were those connected with the division of labor, already foreshadowed by Smith.

In 1832 Charles Babbage, in his work *On the Economy of Machinery and Manufactures*, analyzed the advantages of the division of labor: reduction of the time required to learn a trade, less material wasted on learning, greater skill acquired by repeating the same operations, and greater attention to the manner of using the tools and to improvement in working methods. In this sense, Babbage can even be considered a precursor of industrial engineering.

Another important contribution in this direction was that of Andre Ure, a physician whose scholarly interests included chemistry and physics and a member of the Royal Society of London. In *The Philosophy of Manufactures*, dating from 1835, he conducted an in-depth examination of machines and the mechanization of the production process, as well as the behavior of individual workers and of the labor force as a whole. Ure regarded the "factory system" as the overall set of the "system of machines" and of the "system composed of the workers." In his view, the greatest difficulty posed by the transition to the new systems did not reside in the invention of mechanisms each considered independently, but in their integration into "a single body" and in training workers to adopt new working routines.

These contributions should be considered as premises allowing organization science to evolve as a science in its own right. This would later come about both by virtue of the American Frederick Winslow Taylor (1856–1915), who imposed the revolution of labor organization in the industrial factory, and also by virtue of the French scholar Henri Fayol (1841–1935), who oriented his studies on management of the enterprise as a set of complex functions. These two approaches, both of which marked the beginning of a revolution, were to a large extent complementary and their historical reference to economic knowledge can, as far as Taylor is concerned, be traced back directly to Smith, while the background on which Fayol drew was the conception of the entrepreneur or "undertaker" suggested by Cantillon. It is curious, in this regard, that Fayol focused his studies on the general core management of the enterprise, i.e., on the hierarchy, while Taylor took a more specific interest in the production process, i.e., in the area today defined as "the internal transactional market."

Adam Smith's interpretation of labor as a specialization determined by its division is also present in Taylor's theories. Taylor proposed the scientific organization of labor through careful planning of tasks, thus splitting up the actions and rigidly planning the way they were to be carried out.

What resulted was thus a specialist organization, with the manner of execution overseen by experienced persons in charge of various sectors, who exercised control by measuring performance against pre-assigned times and methods. It was also a form of organization that could be expressed through the chain of command, so that the scope of control could be contained by maintaining direct contact with a maximum of six persons at a time, as had been theorized by V.A. Graicunas in his 1933 work.[23] Finally, this form of organization was based on highly simplified training with obsessive splitting up of tasks, implying an absence of culture in the protagonists of the organizational phenomenon. Such protagonists were demoted to the level of mere executors of what was explicitly imposed upon them and delegated to their immediate execution.

Frederick W. Taylor expounded his theories in two major works, the first dating from 1903, *Shop Management*, and the second from 1911, *The Principles of Scientific Management*. The theories contained in these works, so revolutionary at the time, were given practical application in the manufacturing factories set up by Henry Ford (1863–1947). These factories were designed to carry out mass production of standardized, simple, and often anonymous products, to be imposed on the market with emphasis on quantity rather than quality. These were factories where the predominant approach was that of *make*, i.e., of making as much as possible within the factory, in search of economies of scale to be achieved by reducing to a minimum any purchases from the outside.

For Cantillon's undertaker, on the other hand, who was exposed to the risk inherent in the future, Henry Fayol's studies offered a means of devising tools to minimize this risk. In *Administration industrielle et générale,* published in 1925, the French author analyzed the main entrepreneurial "operations" or functions, dwelling in particular on the principles of governance.

Above all, according to Henri Fayol, "All the operations that are carried out within the enterprise can be divided into the following six groups: (1) *technical* operations (production, manufacturing, transformation); (2) *commercial* operations (purchasing, sales, exchange); (3) *financial* operations (search for and management of capital); (4) *security* operations (protection of goods and persons); (5) *accountancy* operations (inventories, balance sheets, costs, statistics, etc.); (6) *governance* operations (planning, organization, command, coordination and control).

Whether the enterprise is simple or complex, small or large, these six groups of operations, or *essential functions*, are always present."[24] He then argues that "the task of drawing up the enterprise's plan of action, setting up its corporate structure, coordinating its efforts, harmonizing its actions"[25] is a task assigned to the governance function, for which he proposes the following definitions: "*Directing* is to be interpreted as planning, organizing, commanding, coordinating and controlling; *Planning*, i.e. scanning the future horizon and drawing up the plan of action: *Organizing*, i.e. setting up the twofold organism, both material and social, of the enterprise itself; *Commanding*, i.e. ensuring proper functioning of the personnel; *Coordinating*, i.e. enhancing cohesion by joining and harmonizing all actions and efforts; *Controlling*, i.e. ensuring that activity proceeds in conformity with the established rules and orders. Understood in this sense, *directing* is neither an exclusive privilege nor a personal prerogative or task of the head or of the leading figures or top management of the enterprise: rather, it is a function which, like the other essential functions, is performed jointly by the head and the members of the social body. The function of *directing* is to be clearly distinguished from the other five essential functions, and should not be confused with *management*. Managing means leading the enterprise towards its goal, trying to make the best use possible of all the resources of which the enterprise can avail itself; it thus also means assuring the proper working of the six essential functions. *Directing* is but one of the six functions that ensure the proper working of the management, but it holds such an important position among the tasks of the leading figures of the enterprise that it may at times appear as if it were their only function."[26]

He then identified the main principles of governance as follows: "The function of directing has as its organ and its tool only the *social body*. Whereas the other functions utilize materials and machinery, the function of directing acts only upon the personnel. The health and good functioning of the social organism depend on a certain number of conditions, which can interchangeably be called principles, laws or rules. I will here adopt the term *principles*, dissociating it, however, from any idea of rigidity. There is nothing either rigid or absolute in matters pertaining to directing: it is all a question of *measure*. (. . .) Composed of tact and experience, measure is one of the main qualities of the leader of an enterprise. The number of principles involved in *directing* has no limit."[27]

The French scholar placed particular emphasis on "some" of the more important principles that characterize the governance function: "(1) specialization of labor; (2) authority; (3) discipline; (4) unity of command; (5) unity of directing; (6) subordination of particular interests to the general interest; (7) remuneration; (8) centralization; (9) hierarchy or chain of command; (10) order; (11) equity; (12) personnel tenure; (13) initiative; (14) cohesion among personnel, i.e. *esprit de corps*."[28]

Some interesting observations in this regard, which are of great topical interest in the present-day scenario, can be found in reference to the principle of "centralization": "(...) this consists in the fact that in every animal or social organism, sensations converge towards the brain (or the directing center) and that orders proceed from the brain (or from the directing center) and thus set in motion all the parts of the organism. Centralization is not in itself either a good or bad system of directing, since it may be adopted or abandoned at will by the leading figures of the enterprise according to the circumstances; it always exists, to a greater or lesser degree. The problem of centralizing or decentralizing is a simple question of measure. What is needed is to find the balance that is favorable to the enterprise. (...) Finding the measure that will give the best overall performance is the problem of centralization and decentralization. Whatever increases the importance of the part carried out by the subordinate entities is decentralization, whatever decreases this importance is centralization."[29]

Elsewhere Fayol takes up again the definition of governance as a "method" for constructing a science of organization: "I have defined the term 'direction' as 'planning, organization, command, coordination and control', and this definition has been unanimously approved [...] It can be stated that empiricism has reigned supreme up until today in the field of directing enterprises. Every head directs in his own way without troubling to find out whether there exist rules that govern this matter. The absence of a doctrine allows free reign to every kind of whim or fancy. It appears that directing has so far excluded the experimental method, as was the case for medicine until Claude Bernard. However, it cannot be doubted that the majority of the developmental contributions that have brought the science of directing an enterprise to its present level follow the same method, namely the Cartesian method, which can be summarized as follows: 'Observing, gathering and classifying facts, interpreting them, carrying out research whenever possible, and deducing from this set of studies a series of rules which, under the impulse of the heads of enterprises, will enter into the common practice of firms (...).' The problem is that of replacing the general practice of empirical direction by a type of positive direction, which will be based on organization, experience and reasoning. By basing the approach on principles, rules and procedures deduced from experience, there is greater possibility of avoiding mistakes in comparison to the risks which may arise from relying purely on subjective sensations. Empirical does not mean bad: a leader's empirical capability may correspond precisely to the needs of the enterprise. Similarly, positive does not mean excellent: there exists the doctrine and there exists the practice. The best rules can be misunderstood and badly applied. A good doctrine is a valuable tool, but one needs to know how to make use of it."[30]

What is worth underlining is Fayol's observation that it is precisely by planning future action, and keeping it constantly under control, that a limit can be placed on the typical risks faced by the entrepreneur. Thus culture, especially in Fayol, reappears as an element of managerial training, with emphasis placed on coordination. Both of these features characterize current organization theories in cases where the enterprise is examined according to transactional and network models.

The revolutions proposed by Taylor and Fayol evolved over time, becoming consolidated through a number of schools of thought which examined their tenets in greater depth and introduced various improvements. The Soviet writer Dzermen M. Gvisiani, in his book *Management: The Soviet Approach* – a sort of compendium of Western organization theories – classified them as follows: classical theories directly linked to Taylor and Fayol with the fundamental contribution of the bureaucratic school of Max Weber (1864–1920); human relations and informal organization linked to the studies conducted by Elton Mayo (1880–1956); and also

Fritz J. Roethlisberger and Chester I. Barnard's theory of cooperative behavior, the empirical school headed by Peter Drucker, Ernest Dale, Alfred P. Sloan, Alfred D. Chandler Jr., and the movement linked to artificial intelligence conceptions and thus to integration between information and decision, which gave rise to the systemic school.[31] It was in fact the studies on systems theory that led to the analysis of decision-making processes, typical of the thought of the Nobel Prize winner Herbert Simon.

Later, during the 1950s the explorative debate by Stafford Beer and the artificial intelligence school at the Tavistock Institute in London led to the founding of the "sociotechnical" school, where the work by Eric Trist, F.E. Emery, Andrew K. Rice, Robert Kahan, Daniel Katz, Joan Woodward, and many others would come to occupy a prominent role. One of their major concerns was to analyze corporate tools and processes in order to examine their interdependencies with the human factor and to evaluate the influence of technical, social, and economic factors on the human factor itself. A further important aspect of this school was its interest in the use of the computer within the corporate framework. In this sense, they heralded the great IT revolution that would radically modify organizational theory and practice with the introduction of the network enterprise concept.

Ernest Dale, in his 1969 work entitled *The Great Organizers*, simplified the classification of organizational theories by proposing the "universalistic" approach linked to authors such as Taylor and Fayol, who sought to identify general principles and concepts, and the "comparative method" typical of Mayo and various others, who endeavored to draw attention to fundamental similarities among different forms of organization.[32] More recently, a historical school has taken shape, which reexamines organizational theories in the light of the experience acquired over time by large enterprises. Alfred D. Chandler, the most renowned exponent of this school and the author of a number of seminal works, has focused on the relation between structure and strategy. Taking up again the path charted by Adam Smith, Chandler proposes the concept of the *visible hand*, the visible hand of managers and thus of enterprises – a hand which is increasingly becoming predominant as compared to the Smithian *invisible hand*, the latter being typical of the market.[33] In his books Chandler offers several exemplary organizational clarifications, for example, when he defines the meaning of strategy and structure: "The thesis that different organizational forms result from different types of growth can be stated more precisely if the planning and carrying out of such growth is considered a strategy, and the organization devised to administer these enlarged activities and resources, a structure."[34]

Basically, then, starting out from Taylor and Fayol, organization has, at different times, been seen as a mechanism (clock) to be set up with specific procedures and assignments of responsibilities, as a set of types of behavior to be stimulated and controlled, or as something resembling a sociotechnical system with regard to which the information process best suited to decision-making can be established, so that the effects of decisions can be kept under constant control.[35]

Organization theories were subsequently enriched by studies aiming to devise operative management tools. These were seen as consisting mainly of quantitative methods and techniques linked to operative research or springing from the various interpretations of planning in the administrative framework (budget and so forth) and in the sphere of production (sequential planning methods such as PERT, CPM, etc.)

One point, however, is worth underlining. The attention of experts has almost always centered on large entrepreneurial concerns, thus neglecting small- and medium-sized enterprises even though these constitute the basis on which advanced economic systems are founded. An important exception is the work of Alfred Marshall (1842–1924), the British economist whose publications included, in 1879, *Economics of Industry* and, in 1920, the celebrated treatise on the *Principles of Economics*.[36]

Marshall addresses the problems of the entrepreneur, his profits, competition in general, competition between large and small enterprises, and their organization. He was also the first to identify the industrial district as an economic unit of investigation. In Marshall's approach, the district is composed mainly of small- and medium-sized enterprises that have many points in common and which are integrated by rather highly developed forms of competition or cooperation. Today this issue is of great current interest and studies in this area are still one of the important aspects of schools of economic thought, the most significant of these in Italy being that of Giacomo Becattini.[37]

The evolutionary phase of organizational theories, originally ushered in by the industrial revolution that sprang from Adam Smith and was later interpreted by Taylor and Fayol and many other thinkers, continued until the mid-20th century, i.e., up until the advent of the information technology revolution and the rise of advanced technology. The technology revolution signaled a dramatic solution of continuity that would completely change the rules of the game, including rules concerning the organization and management of enterprises.

Once again, it was to be an economic thinker with a background in law, as had been the case with Adam Smith, who would prefigure the emerging problems, the full complexity of which was not fully grasped until the last decades of the 20th century. This towering figure was that of Ronald Coase, an American economist of British origin, born in London in 1910.

In 1937, in an article that would become famous only many years later, Coase laid the basis for a new analysis in *The Nature of the Firm*.[38] As a professor of economics at the University of Chicago, he was awarded the Nobel Prize in 1991 for his discovery of the role of transaction costs and property rights in the modeling of institutional structure. The concept of transaction is defined by Coase in the following manner: "in order to carry out a market transaction it is necessary to discover who it is that one wishes to deal with, to inform people that one wishes to deal and on what terms, to conduct negotiations leading up to a bargain, to draw up the contract, to undertake the inspection needed to make sure that the terms of the contract are being observed, and so on."[39] Coase then proceeds to specify the various costs involved: "Dahlman crystallized the concept of transaction costs by describing them as 'search and information costs, bargaining and decision costs, policing and enforcement costs.' "[40] Coase took these costs as the basic framework for the concept of the size of the enterprise: "the upper limit on the size of the firm is reached where the costs of organizing a transaction equal the cost of making that transaction in the market."[41]

The concept of the transaction as the unit of investigation and analysis of organizations – and the related issue of costs, which constituted a genuine new revolution – would rise to prominence theoretically at the moment when information technology and the new field of information and communications technology began to be developed.[42]

According to Francesco Silva, who inquired into the nature of the enterprise, Coase "examines the aspect of transaction costs, whose existence explains the existence of the enterprise itself, as an organizational-contractual structure that allows a saving on such costs, although at the same time generating other costs of a managerial type [. . .] In actual fact, however, it was not until the 1970s, in particular not until the contributions by O. Williamson and A. Alchian-H. Demsetz, that the idea of transaction costs became a constitutive element of a new paradigm of industrial organization theory, and more in general of economic theory. This gave rise to the theory of the firm, and a fresh look was taken at the theory of production integration. Economists began to look more carefully at the problem of economic organization and thus also inquired more closely into their institutional setting."[43]

The analysis of transaction costs leads to a reexamination of the concepts of hierarchy and market within the enterprise. These are concepts that Oliver E. Williamson addressed above all in *Market and Hierarchies: Analysis and Antitrust Implications*, published in 1975. In Coase's view, "Williamson's alternative approach was built up on the distinction between markets and hierarchies, it incorporated transaction costs into the analysis, and examined the problems of the internal organization of the firm in far greater detail than is usually the case with economists. This approach is largely consistent with *The Nature of the Firm* [. . .]."[44]

The groundwork was thus laid for the theoretical basis of a revolution in organizational theory, which was to go hand in hand with the IT revolution. The implications of this new setup were to be perhaps even more momentous than the outcome of the industrial revolution itself. The use of machines based on information technology (computers and robots) spread rapidly throughout the world of the factory and the enterprise, establishing an indissoluble bond with the operators. Thus there arose new forms of individuality, namely, the pair composed of individual and computer. The effect was to recompose – to use a metaphor by Federico Butera – the shattered remnants of the fragmented and repetitive Tayloristic pattern, which was now entrusted to robots.[45] Furthermore, culture once more appeared as an indispensable element enabling the individual to guide "smart" machines even if the relation of technological assignment of responsibility between human being and computer opens up the risk of an intellectual neo-Taylorism, especially in some new forms of *knowledge management*.

The picture that thus emerges is one of the rediscovery of the human factor, with the result that professional training is called upon to assume a new role more explicitly linked to education and culture. This entails a substantial change in the world of the enterprise, which is now strongly influenced by the intricate workings of IT machinery that is capable of generating a network of transactions not only with regard to administrative management of the enterprise as a whole but also in the factory, thus establishing the groundwork for a new vision of the organizational phenomenon.

These considerations lead in to the concept of the transactional enterprise. In addition, terms such as the short, lean, or streamlined enterprise have been introduced, in reference to the endeavor to engage in drastic pruning of vertical connections in organization charts. A further recent development concerns the model of the network enterprise, whose mesh of interactions is capable of interpreting both hierarchical relations and also the web of transactions occurring among the nodes or organizational poles.

Seen in this light, the enterprise now appears as a complex network of generally new technologies, mainly Information and Communication Technologies (ICT), to which there corresponds a network of highly educated individuals – knowledge workers – who are capable of operating on the IT machines almost as if they were self-entrepreneurs. The resulting model is reminiscent of the thought of the economist Joseph A. Schumpeter, since knowledge workers effectively generate a constant stream of innovation that also makes use of inputs emanating from the organizational form embodied by the macroenterprise. The quest for flexibility is thereby facilitated, with priority being awarded to the *buy* logic– of purchasing mainly from external sources – as compared to the *make* approach, which would instead concentrate on making as much as possible within the factory itself.

Thus the structure truly becomes a governed network of hierarchical interactions and of internal and external transactions occurring within the macroenterprise. The internal market merges with the hierarchy, the latter tending to become flat, so that interpersonal relations likewise assume the form of a network. But in this perspective, governance becomes an essential function for organizational coordination, and knowledge becomes a strategic element of governance that helps to prevent the dangers of centrifugal forces, i.e., the risk that the increasing independence of the parts may result in disintegration of the overarching entrepreneurial unit. Modifications are also gradually introduced in the role of the administrative functions, with greater significance being awarded to sourcing and supplies (purchasing function), which act as a key to governing the configuration of the macroenterprise.

These are the conditions that are necessary in order to achieve the present-day streamlined form of production based on lean supply, to use the definition offered by Romano and Danese, namely " lean supply concerns evaluation of the customer-supplier relation rather than that of the individual supplier," or again "Lean subsupplying [. . .] entails that suppliers all have to develop their technologies independently of the assemblers' requests, with the result that each of the suppliers can become a leading enterprise in technology, a genuine innovator and a driving partner. [. . .] Moreover, even the customer is called upon to share information with the suppliers, for example information on processes, including data on costs, or on sales plans, and to accept the suggestions and ideas arriving from upstream. In this case, the customer and the supplier are truly placed on the same plane."[46]

Notes

1. *Grande Dizionario della lingua Italiana* [*Complete Dictionary of the Italian Language*], vol. XVII, UTET, Turin, 1997, p. 32.
2. Charles Davenant as quoted by Joseph A. Schumpeter in *History of Economic Analysis*, Allen & Unwin, London, 1954, p. 76.
3. Fritjof Capra, *The Hidden Connections: A Science for Sustainable Living*, Doubleday, New York, 2002.
4. Franco Venturi, *Introduzione* a Cesare Beccaria, *Dei delitti e delle pene* [*Introduction* to Cesare Beccarla, *On Crimes and Punishments*], Turin, Einaudi, 1973, vii.
5. Thomas S. Kuhn, 1962, *The Structure of Scientific Revolutions*, University of Chicago Press, Chicago, 1962, p. 10.
6. Ibidem, p. 6.
7. Ibidem, p. 42.
8. George Duby, *L'Histoire continue* [*History Continues*, Chicago, University of Chicago Press, 1994], Odile Jacob, Paris, 1991, p. 8.
9. René Thom, *Stabilità strutturale e morfogenesi. Saggio di una teoria generale dei modelli* [*Structural Stability and Morphogenesis*, New York, Perseus Books, 1999], Einaudi, Turin, 1980, p. 50 ff.
10. Eric Hobsbawm, *The Age of Extremes: The Short 20th Century, 1914–1991*, Michael Joseph, London, 1994.
11. Piero Ottone, *Potere economico* [*Economic Power*], Longanesi, Milan, 1968, p. 44.
12. Ibidem, p. 48.
13. Richard Cantillon, *Essai sur la nature du Commerce en Gènèral traduit de l'Anglois*, A Londres chez Fletcher Gyles dans Holborn, M.DCC.LV.; in English, *Essay on the Nature of Commerce in General,* English translation by Henry Higgs 1931; with a new introduction by Anthony Brewer, Transaction Publishers, New Brunswick, NJ, 2001, pp. 23–25.
14. Frank H. Knight, *Risk, Uncertainty, and Profit*, Houghton Mifflin Company, Boston, MA, 1921.
15. Alexander Bogdanov, *Tektologiya: Vseobschaya Organizatsionnaya Nauka*, in 3 volumes, Berlin and Petrograd-Moscow, 1922. English translation as *Essays in Tektology: The General Science of Organization*, trans. George Gorelik, Intersystems Publications, Seaside, CA, 1980. In Italian, for a summary cf. Aleksandr Alexandrovïc Bogdanov, *Saggi di scienza dell'organizzazione* [*Essays in Organization Science*], Theoria, Rome, 1988.
16. Piero Ottone, *Potere economico* [*Economic Power*], cit., p. 49.
17. Ibidem, p. 51.
18. A. Roncaglia and P. Sylos Labini, *Il pensiero economico. Temi e protagonisti* [*The Wealth of Ideas: A History of Economic Thought*, Cambridge University Press, 2005], Laterza, Bari, 1995, pp. 16–17.
19. William J. Barber, *A History of Economic Thought*, Penguin, New York, 1967, p. 10.
20. Lionel Robbins, *La misura del mondo. Breve storia del pensiero economico* [A History of Economic Thought: *The LSE Lectures*, edited by Warren J. Samuels and Steven G. Medema, 1998, http://ebooks.ebookmall.com/ebook/116947-ebook.htm], Ponte delle Grazie, Milan, 2001, p. 195.
21. William J. Barber, *A History of Economic Thought*, cit., Prologue.
22. Adam Smith, *An Enquiry into the Nature and Causes of the Wealth of Nations*, Straham & Cadell, London, 1776, http://www.adamsmith.org/smith/won-b1-c1.htm.
23. V.A. Graicunas, *Relationship in Organisation,* first published in *Papers of the Science of Administration*, edited by L. Gulick and L.F. Urwick, Institute of Public Administration, New York, 1937.
24. Henri Fayol, *Administration industrielle et générale: prévoyance, organization, commandement, coordination, contrôle* [*Industrial and General Administration,* Translated from French by J. A. Coubrough, Pitman, 1930], Dunod, Paris, 1925, p. 23.

25. Ibidem, p. 26.
26. Ibidem, pp. 26–27.
27. Ibidem, p. 43.
28. Ibidem, p. 44.
29. Ibidem, pp. 59–60.
30. Henri Fayol, "L'administration positive dans l'industrie" ["Positive Administration in Industry"] in *L'éveil de l'esprit public* [*Awakening the Public Spirit*], Dunod, Paris, 1927, pp. 267–275.
31. Dzermen M. Gvisiani, 1970, *Management e l'approccio sovietico* [*Management and the Soviet Approach*], Etas, Milan, 1971.
32. Ernest Dale, *The Great Organizers,* McGraw-Hill, New York, 1960.
33. Alfred D. Chandler, *The Visible Hand: The Managerial Revolution in American Business*, The Belknap Press, Harvard, Cambridge, MA, 1977.
34. Alfred D. Chandler Jr., *Strategy and Structure: Chapters in the History of the Industrial Enterprise*, MIT Press, Cambridge, MA, 1962, p. 13.
35. Federico Butera, *L'Orologio e l'Organismo – Il cambiamento organizzativo nella grande impresa in Italia: cultura industriale, conflitto, adattamento e nuove tecnologie* [*The Clock and the Organism – Organizational Change in the Large Enterprise in Italy: Industrial Culture, Adaptation and New Technology*], Franco Angeli, Milan, 1984.
36. Alfred Marshall, *Principles of Economics*, Macmillan and Co., Ltd., London, 1920, Eighth edition, First published 1890.
37. In particular cf. *Economia senza gabbie. Studi in onore di Giacomo Beccatini* [*Economics Without Cages: Studies in Honor of Giacomo Becattini*], edited by Nicolò Bellanca, Marco Dardi and Tiziano Raffaelli, Il Mulino, Bologna, 2004.
38. Ronald H. Coase, "The Nature of the Firm," in *Economica*, 1937, pp. 386–405.
39. Ronald H. Coase, *The Firm, the Market and the Law*, University of Chicago Press, Chicago, 1988, p. 6.
40. Ibidem, p. 6.
41. Ibidem, p. 7.
42. Ibidem, p. 36.
43. Francesco Silva, Ronald H. Coase and "The Nature of the Firm", in *Economia e Politica Industriale*, n° 70, 1991, Angeli, Milan, p. 251.
44. Ronald H. Coase, *The Firm...* cit., p. 80.
45. Federico Butera, *I frantumi ricomposti* [*The Recomposed Shards*], Marsilio, Venice, 1972.
46. Pietro Romano, Paolo Danese, *Supply Chain Management. La gestione dei processi di fornitura e di distribuzione* ["Managing Business Processes Across Supply Networks: The Role of Coordination Mechanisms," *Journal of Purchasing and Supply Management*, digital document, Elsevier 2004], http://www.sciencedirect.com/science?_ob=ArticleURL&_udi=B7579-4FBHW7M-2&_user=10&_rdoc=1&_fmt=&_orig=search&_sort=d&_docanchor=&view=c&_searchStrId=945307772&_rerunOrigin=google&_acct=C000050221&_version=1&_urlVersion=0&_userid=10&md5=bb2baa539edff152b59783e6f1cc83e6,], McGraw-Hill, Milan, 2006, p. 123.

Chapter 5
The Production Pole: From Fordism to Toyotism

A glance at the history of entrepreneurial organization reveals that the most significant developments have certainly been those occurring in the production pole. Here, the changes that have taken place over time have often affected the subsequent pattern of the entire corporate management.

A useful starting point for a more detailed examination of organizational change in the history of the production function – i.e., of the production pole, seen in terms of the network enterprise – is the picture emerging from manufacturing industry, with special reference to the automotive industry. "Forty years ago Peter Drucker dubbed it 'the industry of industries'. Today automobile manufacturing is still the world's largest manufacturing activity, with nearly 50 million new vehicles produced each year."[1] But this is by no means the only justification for such a phrase. First, the process and the organizational changes introduced into the automotive sector rapidly spread throughout the major industrial activities. Second, another salient aspect of this evolutionary path is that the two most important revolutions in the organization of production processes were imagined and indeed more or less completely brought into being by the joint effort of four great personalities, who succeeded in combining theory with practical application. First in this line of historic names is the towering figure of Frederick W. Taylor, who, together with Henry Ford, proposed the scientific organization of labor in the factory environment. This approach would soon become known as "Taylorism–Fordism" and was destined to become the classic example of mass production.

In opposition – but partly as a complement – to this approach, a new conception of industrial labor began to take shape in the Toyota production plants in Japan during the second half of the 20th century. The new Japanese approach was given various names, "Toyota system or spirit," "integrated factory," "Toyotism," and more generally "flexible" or "lean" production. This revolutionary way of conceiving the production process was the brainchild of a Japanese technician, Taiichi Ohno, born in 1912 in Manchuria, who theorized and devised his own version of a production system (Toyota Production System) inspired by the Taylor model. The Ohno version was then implemented in the Toyota factories by the legendary president of the company, Kiichiro Toyoda. The innovation proved so revolutionary that it induced several US scholars to entitle their research on the topic as "The machine that changed the world."

G. Dioguardi, *Network Enterprises*, Innovation, Technology, and Knowledge Management 104, DOI 10.1007/978-1-4419-1333-3_5, © Gianfranco Dioguardi, 2010

The organizational changes represented by the Japanese production system were substantial, but the most important aspect consisted of the radical change in production strategies, which in some respects were more or less inverted as far as the traditional process flow was concerned. What changed, effectively, was the classical logistic chain based on the conception of "Production–Delivery–Sales," which thinkers such as Pierre Coureil indicated as the "product exit" production model (push method). In the new system, priority was awarded to effective demand, thereby imposing the chain "Sales–Production–Delivery" in a "client entry" service model (pull method).[2]

In the traditional production system, the first step consists in establishing the plans that will form the basis for setting up the plants and establishing the volumes of production. This is followed by proceeding from upstream to downstream with the flow of components to be assembled, until what is achieved is the finished product, which is effectively mass produced and in a sense "imposed" on the final recipient. The latter is regarded as a passive purchaser, and as the expression of a generally rather widespread demand, who for this very reason is inclined to purchase a product without delving too deeply into its quality.

Clearly, a system of this kind is rather rigid, and amortization of fixed installations can be achieved only by means of plans extending over the medium and long term. There is no scope for any variation when the quality has been designed and planned for an entire series of products.

The Toyota system, on the other hand, embodies a radically different conception, tantamount to a diametrically opposed outlook, to the point that Taiichi Ohno defines it as "thinking in reverse." Ohno himself explains its philosophy as follows: "In automobile production, material is machined into a part, the part is then assembled with others into a unit part, and this flows towards the final assembly line. The material progresses from the earlier processes toward the later ones, forming the body of the car.

Let's look at this production flow in reverse: a later process goes to an earlier process to pick up only the right part in the quantity needed at the exact time needed. [...] Then the method of transferring the materials is reversed. To supply parts used in assembly, a later process goes to an earlier process to withdraw only the number of parts needed,"[3] or again "a later process goes to an earlier process to withdraw parts needed just in time... to make certain the earlier process produces only as many parts as are picked up by the later process."[4]

The Japanese technician further notes: "In short, where the Ford system sticks to the idea of making a quantity of the same unit at one time, the Toyota system synchronizes production of each unit. The idea behind this approach is that in the marketplace, each customer buys a different car, so in manufacturing, cars should be made one at a time. Even at the stage of making parts, production is carried out one piece at a time."[5]

What begins to come to the fore here, then, is the figure of the recipient to whom the product is destined. Recipients are now seen as customer-clients seeking achievement of the quality characteristics they desire. At the same time, warehouses for stocking of final as well as intermediate products are eliminated because

"At Toyota, production quantity equals market demand or actual orders. In other words, the number needed is the number sold. Therefore, because market needs are directly connection to production, manufacturing cannot arbitrarily change production quantities. To reduce overproduction, efficiency improvement must be achieved on the basis of required numbers. In other words, production quantities are based on demand."[6] Similarly, "[r]equired numbers are based on sales and this is determined by the market. Consequently, production is given a number based on demand or the actual orders...."[7]

It follows that one of the hallmarks of the Toyota system is its great flexibility, allowing a remarkably lean production factory, where the concept of total quality is applied in order to adapt production to the real individual requirements of customers. However, marketing and market research still represent important aspects of the overall system, partly as a means of conducting a preliminary investigation on the market, but also to establish general guidelines for a type of production that will necessarily be rendered flexible by the inverted downstream to upstream pathway.

Let us now take a closer look at the details of these two systems. At the beginning of the 20th century, Fordism–Taylorism, i.e., "scientific labor organization," gained widespread acceptance, starting from the Ford production plants, where manufacturing industry was developing mass production with the first standardized cars. It was in these factories that the epoch-making transition from individualized artisanal production to the rigidly structured assembly line took shape. On the assembly line, production was rigidly planned, with products moving along the assembly line while the workers would remain more or less stationary in their positions, merely inserting the standard pieces according to the preplanned operations as each product passed in front of them. This was argued to allow more rational assembly of the different components. Ohno provides an eloquent account of this phenomenon in his account of the Ford system: "This is the real mass production system in which raw materials are machined and carried along conveyer belts to be transformed into assembled parts. The components of various types are then supplied to each of the final assembly processes, with the assembly line itself moving at a fixed speed, as parts are assembled to ultimately become fully assembled cars coming off the line one by one."[8]

The transition from craftwork to mass production was indeed a dramatic turning point, which introduced a ground-breaking change. According to Mario Deaglio, "Lean production is identified as the third mode of production developed by industrial societies, after the original mode – which was based on craftwork – and after mass production. Craftwork production was characterized by the elevated demand for highly skilled manpower, technically simple but flexibly usable physical capital, and production at extremely high cost carried out at the consumer's request. In mass production, on the other hand, the majority of the workers are fairly unskilled whereas enterprises are well equipped with fairly high level physical technological capital. In mass production processes, the machines almost always perform only one function, resulting in loss of the flexibility that characterized craft production [...] Lean production can [...] be defined as the highly positive synthesis, the sum of the best features of the two earlier production modes, the extremely successful blend of craftsmanship and the low cost of mass production. Absence of defects and low costs: this is the secret of Japanese success."[9]

These basic concepts have remained virtually unchanged since then, with the exception of occasional modifications, above all in Volvo's factories, where European automobile manufacturers conducted a few marginal experiments, with the participation of the workers. For instance in Volvo's Kalmar factory, experiments were conducted with a sort of revival of Henry Ford's 1910 assembly plant, reintroducing the craftwork techniques by assigning the assembly of a whole vehicle to small groups of workers.[10] This issue was also taken up by Taiichi Ohno, who stated that "Even though today some manufacturers – Volvo for instance – have one person assemble the entire engine, in general the manufacturing mainstream still utilizes Ford's work flow, or automation, system. [. . .] The difference is that, while Sorenson worried about warehousing parts, Toyota eliminated the warehouse.[11] Therefore, the right parts needed in assembly reach the assembly line at the time they are needed and only in the amount needed."[12]

In Ford factories the entire production was carried out at the plant, with the major emphasis being placed on what was defined as the *make* approach, i.e., doing everything in-house. Individual parts may have been produced in places that could be likened to independent workshops dealing with specialized productions, but they were still contained within the overall plant. This situation is well depicted in the following overview by the American researchers Womack, Jones, and Roos: "By 1915 [...] Ford had taken all these functions in-house and was well on his way to achieving complete vertical integration (that is, making everything connected with his cars fro m the basic raw materials on up) [. . .] Ford . . . could achieve substantial cost savings by doing everything himself. [. . .] So he decided to replace the mechanism of the market with the 'visible hand' of organizational coordination."[13]

The question of parts constituted a serious issue, for in the typical model of an automobile, which is composed of over 10,000 parts, the designing and building of individual parts certainly represents the major challenge. As the above-cited authors further point out, "Henry Ford thought he had solved the problem by the time of World War I: Do it all yourself in your own company. However, his solution raised at least as many questions as it answered: How do you organize and coordinate thousands of employees in hundreds of factories and engineering offices? [. . .] In the 1920s Alfred Sloan found one answer to these problems: Do it all in your own company, but set up decentralized parts-making divisions as independent profit centers [. . .].By treating these divisions as independent businesses Sloan thought he could impose the cost and efficiency discipline of the market while preserving the coordination advantages of a unified company."[14]

Alfred P. Sloan was the legendary manager-business executive who devised an organizational revolution that brought General Motors to the peak of success. Sloan described his idea in a famous autobiographical book – *My Years with General Motors* – in which he also proposed a variety of models of automobiles to be offered to the general public. This was conceived as a means of addressing the problem of market saturation he was beginning to perceive.[15] Ohno has the following recollections of the great manager: "Sloan, [who was the Chairman of the Board of General

Motors], grabbed hold of this important modification in the marketplace and began to offer more and more different models [...] In this transition from mass-produced Model Ts to the full-line policy of General Moors, production processes became complicated."[16]

As we have seen, Ford endeavored to improve the economic process through the *make* conception, i.e., the idea of producing everything in-house, so that the costs to be borne by the product would be represented exclusively by internal costs without any surcharges attributable to profits siphoned off by external enterprises. However, the negative side of this state of affairs was that in order to develop the entire production in-house, extremely large plants had to be built, and this called for substantial investments that would necessarily involve a medium- and long-term perspective. Thus the system was almost devoid of in-built flexibility: indeed, the very fact of having fixed installations entailed a further element of rigidity, since these were adapted to a specific product which could not easily be changed once production had been set up. Therefore, the production process had to be rigidly planned for a period of time that would allow the fixed installations to be amortized. During this period, no variation in production could be introduced and production itself would have to remain unchanged in conformity with the previously established plans.

This necessarily unchanging and long-term perspective on production made it possible to apply standardized and repetitive methods that reflected the scientific organization of labor as conceived and studied by Taylor. The entire production process was split up into individual tasks in which operations could be carried out through simple movements, choosing the most effective and most efficient procedure (the *one best way*) for each action. For each specific task a standard length of time required to perform the action was calculated, and this was the goal to be pursued by each worker performing the task, after the worker had been given preliminary instructions as to the best way to achieve the goal most efficiently. Workers were to concentrate exclusively on manual abilities, developing dexterity in the most effective manner. This type of organization was also encouraged by economic incentives which gave rise to various forms of piece rate work, with performance-related pay. The resulting setup made it possible to impose even greater subdivision of labor, with extremely pronounced – even obsessive – fragmentation of tasks, with the aim of achieving greater and greater standardization.

This situation was not without its adverse repercussions on the social level, leading at times to serious social conflict that was only partly mitigated by successive attempts to improve working conditions through forms of job enlargement, or job enrichment, or sometimes job rotation, or other types of improvements over the original working methods.[17] But the fact remained that in its fundamental structure, Taylorism was based on task management, i.e., on an organizational system which involved the assignment of pre-established and well-defined tasks. This was done by breaking down the activities into a preliminary planning process entrusted to a purpose-appointed office and then the more specific stage of task execution, which called for elevated standardization as well as extreme fragmentation of tasks into particularized sub-tasks. Manual labor was thus split up and subdivided into elementary, highly simplified, and standardizable tasks which the workers were obliged to perform according to the prescribed mode of action, since the widespread standardization did not concern only the equipment and the pieces to be worked on but also the actual ways in which the tasks were executed in the factory environment. Furthermore, workers were required to carry out their tasks in an extremely repetitive manner, with speed of action in task execution being the prime consideration: indeed, so important was speed that execution was constantly monitored by specific control units that were in charge of task execution methods and timing and patrolled the factory in order to supervise efficiency of performance. Charlie Chaplin gave a brilliant illustration, in the motion picture *Modern Times*, which showed that while such a system was intended as a quest for extreme efficiency of production, it was unsustainable for the human being. Taylor himself hypothesized that all intellectual activity should be eliminated from the factory workplace and should instead be concentrated in the planning office. The management, for its part, would supervise and monitor task execution "by exceptions," focusing attention only on any activities and results that strayed from the planned and expected outcomes.

Since speed and standardization of action were of paramount importance, overseers and foremen were assigned strictly executive tasks. According to an insightful scholar of organizational theories, Aldo Fabris, the Tayloristic model thus presented the following features: "The operative elements, not always clearly and organically described, can be ordered as follows: (a) functional management with responsibilities clearly divided among various persons and establishment of a planning center or office; (b) standardization of all tools, operations and movements for all aspects of labor, utilization of the exception principle and application of modern costing calculations; (c) introduction of the concept of clearly defined task and application of differentiated piece rates; (d) utilization of appropriate monitoring and control tools (slide-rules, timing devices, instruction sheets, mnemonic classification systems, charts for the flow of materials)."[18]

Taylor's organizational model was also designed to be implemented in rigidly functional terms. Fabris again offers insight into this aspect: "Traditional organizational structures [...] were [according to Taylor] of a 'hierarchical-military' type, in other words they followed a linear pattern; by contrast, one of the necessary conditions for application of the new system is a function-based type of organization which consists in subdividing organizational labor in such a way that each man, from the superintendent down, has the lowest number of functions to 'perform.' "[19]

The organizational models expressed by the functional organization charts defined the lines of connection between the corporate management and the various operative tasks carried out in the enterprise. Among the different functions, which were initially analyzed and studied in depth by the French engineer Henri Fayol, it is of relevance to mention sourcing and purchasing, production, sales, personnel, organization, finance, administration, management control, research and development, information systems, and marketing.

For this type of organization, the shop foreman was ultimately under the supervision of eight functional overseers whose supervisory activities partly involved the planning office and partly the actual factory floor itself. In the work by Aldo Fabris they are identified in the following manner: "*In the planning office*: (1) order clerk-cum-production worker (who establishes the exact route each piece has to take through the factory in order to be completed within the appointed time, and writes out the worksheet in order to inform the overseers and the workmen as to the exact succession of tasks); (2) the man in charge of work sheets (the work sheet provides the overseers who have executive tasks and also the workers themselves with all the details of their work and the piece rates); (3) the man in charge of times and costs (who supplies workers with the wherewithal to time their work and record the times, and makes sure that all record sheets are handed in duly filled out by the workers); (4) the man in charge of disciplinary relations (who intervenes in cases of insubordination and takes the necessary measures, and also maintains an updated record of the good and bad performances of the individual workers, which must be consulted prior to an wage variation).

In the factory, with executive tasks: (5) foreman properly speaking (in charge of preparing all the work right up to the moment when a part is mounted onto a machine); (6) speed of execution instructor (his work begins when the part is mounted on the machine, and he is entrusted with making sure that for each part the appropriate tools are used and that all the specific rules are obeyed in all the other aspects of the work); (7) inspector (responsible for quality control and both the workers and the foremen have to follow his prescriptions scrupulously); (8) repair officer (who is in charge of ensuring that all the workers utilize the machines according to his prescriptions, and who takes care of maintenance)."[20]

The production process was initiated after a preliminary action of marketing that was regarded as having predominantly strategic functions. The market investigation was designed to provide forecasts that would answer the classic questions as to what to produce and what was the most appropriate quality for a given product, and how much to produce in relation to the potential market that could most likely be conquered. The next stage then sought to define the market in terms of its size and localization. Steps were also taken to establish the presumed price that should be indicated to potential customers, setting the price on the basis of an analysis of expected internal costs. This analysis was only partly linked to external expectations deriving from the intersection of the demand and supply curves plotted on the basis of statistical data and to expectations springing from forecasting analysis.

Planning strategies were established ex ante with respect to the production process, and they essentially defined the organizational structure of this process. Thus it was the planning strategies that formed the framework and the rationale for designing the size and quality of the installations and the assembly line and for organizing the stocks: all this required a considerable commitment of funds in order to embark on the massive one-off investment. Clearly, therefore, what was created was a rigid production apparatus which, once started up, could no longer be modified. Thus even the final production was often carried out not so much for immediate sale as, rather, to be stockpiled in warehouses where it would be disposed of little by little over time. In effect, the actual sale of the products was to be accomplished by, as it were, "obliging" customers to purchase what had been produced, in a process that was boosted and underpinned by a form of advertising that not infrequently verged on veritable propaganda. It is worth recalling the singular assertion by Henry Ford concerning the fact that his customers were to purchase whatever car they wished as long as it was black and of the celebrated standardized model, denominated "T." The latter was a typical expression of mass production of uniform, unexceptional, and inexpensive commodities. According to the authors of *The Machine That Changed the World*, "Ford's 1908 Model T was his twentieth design over a 5-year period that began with the production of the original Model A in 1903. With his Model T, Ford finally achieved two objectives. He had a car that was designed for manufacture... and that was ... user-friendly. Almost anyone could drive and repair the car without a chauffeur or mechanic. These two achievements laid the groundwork for the revolutionary change in direction for the entire motor-vehicle industry.

The key to mass production wasn't [...] the moving, or continuous, assembly line. *Rather, it was the complete and consistent interchangeability of parts and the simplicity of attaching them to each other.* These were the innovations in production that made the assembly line possible."[21] And in addition "Ford's first efforts to assemble his cars, beginning in 1903, involved setting up assembly stands on which a whole car was built, often by one fitter. [...] The assembler/fitters... had to get the necessary parts, file them down so they would fit (Ford hadn't yet achieved perfect interchangeability of parts [...] Then, around 1908, when Ford finally achieved perfect part interchangeability, he decided that the assembler would perform only a single task and move from vehicle to vehicle around the assembly hall. [...] Ford's stroke of genius in the spring of 1913, at his new Highland Park plant in Detroit, was the introduction of the moving assembly line, which brought the car past the stationary worker. [...] Ford not only perfected interchangeable part, he perfected the interchangeable worker."[22] This also made it much easier to train the operators in a workplace that was by now characterized by extreme job fragmentation.

Consumers thus ended up being slaves of mass production, a phenomenon that was in any case exacerbated by an economy of scarcity, where the need to ensure quantity of production was more greatly felt than any need for attention to product quality. Quality itself was generally checked ex post with respect to the time of production, and it therefore displayed typically destructive characteristics, given that any product which failed to meet the pre-established standards necessarily had to be discarded.

Notes

1. James Womack, Daniel T. Jones, Daniel Roos, *The Machine That Changed the World*, Harper Perennial, New York, 1991, p. 11.
2. Pierre Coureil, *Valore Aggiunto. Costruire e sviluppare le competenze dell'impresa* [*Added Value: Building and Developing the Competences of the Enterprise*], Edizioni Olivares, Milano, 1998, p. 121.
3. Taiichi Ohno, *Toyota Production System*, 1978, English translation copyright by Productivity Press, New York, 1988, p. 5.
4. Ibidem, p. 95.
5. Ibidem, p. 96.
6. Ibidem, p. 61.

7. Ibidem, p. 127.
8. Ibidem, p. 93.
9. Deaglio, *Introduction* to *La macchina che ha cambiato il mondo*, Milan, Rizzoli, 1993 [*Introduction* to Italian translation of Womack et al., *The Machine That Changed the World*], xvi–xvii.
10. Womack et al., *The Machine*, cit., with the discussion of "neocraftsmanship," p. 101.
11. Taiichi Ohno, *Toyota Production System*, cit., p. 95.
12. Ibidem, p. 4.
13. Womack et al., *The Machine*, cit., pp. 33–34.
14. Ibidem, pp. 138–139.
15. Alfred P. Sloan, *My Years with General Motors*, Doubleday, New York, 1963; Italian translation: *La mia General Motors*, edited by John McDonald with Catharine Stevens, Preface by Gianfranco Dioguardi, Il Sole 24 Ore, Milan, 1991.
16. Taiichi Ohno, *Toyota Production System*, cit., p. 104.
17. Giuseppe Bellandi, *Economia e gestione dell'impresa* [*Economy and Management of the Enterprise*], Utet, Turin, 1993, pp. 134–135.
18. Aldo Fabris, *Storia delle teorie organizzative* [*History of Organization Theories*], Isedi-Mondadori, Milan, 1980, p. 42.
19. Aldo Fabris, *L'organizzazione dell'impresa* [*The Organization of the Enterprise*], Etas libri, Milan, 1983, p. 60.
20. Fabris, *Storia* [*History*], cit., pp. 43–44.
21. Womack et al., *The Machine*, cit., p. 26.
22. Ibidem, pp. 27–30.

Chapter 6
The Toyota Spirit

The underlying philosophy characterizing the Toyota spirit can indeed be described as "thinking backward," in the sense of "establishing, in parallel with the development of the real production flow (which proceeds from upstream to downstream), an inverse flow of information which proceeds from downstream to upstream, since each downstream location sends an instruction to the next higher location [. . .] The innovation, as can be seen, is purely organizational and conceptual: there is no intervention of any technological aspect."[1]

In Japan, Toyota first embarked on the process of change in the 1970s and continued to introduce further changes for over a decade. Taiichi Ohno points out that "[...] America had cut costs by mass-producing fewer types of cars. It was an American work style – but not a Japanese one. Our problem was how to cut costs while producing small numbers of many types of cars."[2] What this suggests is that the Japanese approach to production had felt the need to adapt to a market that was profoundly different from the original circumstances in which Ford had set up its factories. What was necessary now was not simply to produce, but above all to actually sell the products that had been manufactured in order to avert the financial insolvency typical of overproduction, as frequently occurred in the United States during the repeated crises whose origins are eloquently depicted by Guido Rossi: "When banks enter onto the scene – immediately becoming shareholders of the enterprises – the result is a rampant and uncontrolled growth of the system, which leads to repeated crises, the 1929 crisis being just the paradigmatic example of the innumerable cases. The mechanisms that triggers them has a perverse simplicity of its own, and the script is repeated over and over again: enterprises produce far more than is required, and their goods remain unsold. Together with the rising accumulation of stocks and inventories in warehouses, their indebtedness towards the banks also increases, and precisely because of the collapse of sales the enterprises cannot, or can only partly, honor their debts. At this point the banks, keen to avert insolvency and company bankruptcies, maintain the share prices artificially high by purchasing shares themselves or persuading their customers to do so."[3]

At this juncture the system, in particular the financial system – but consequently the economic system as well – becomes dangerously over-extended and endures only as long as it can withstand the smoldering flames of crisis. And when the bubble finally bursts, the upshot for an enterprise is disastrous. This prompts a series of

G. Dioguardi, *Network Enterprises*, Innovation, Technology, and Knowledge Management 104, DOI 10.1007/978-1-4419-1333-3_6, © Gianfranco Dioguardi, 2010

searching questions, which Taiichi Ohno raises in his important book: "[...] What are the essential needs of business under slow growth conditions? In other words, how can we raise productivity when the production quantity is not increasing? [. . .] There is no waste in business more terrible than overproduction. Why does it occur?"[4]

The most natural answer stems from the realization that it is imperative to start out from real needs, giving up the idea of accumulation of stocks: "Industrial society must develop the courage, or rather the common sense, to procure only what is needed when it is needed and in the amount needed."[5] This explains the introduction of just in time and above all of the so-called *kanban* method: "The first rule of *kanban* is that the later process goes to the earlier process to pick up products. This rule was derived from need and from looking at things upside-down, or from the opposite standpoint."[6]

These more detailed observations on the way the Japanese production model is structured help to highlight some substantial changes affecting both the production system and also the sales-related market activities. In fact, the Japanese approach paints a very different picture even where the actual conception of the customer is concerned, with the customer transformed into nothing short of a veritable client who commissions the product that will be purchased. This is a crucial aspect, which is of particular relevance "on markets that are close to saturation and are ever more exacting, ever more selective. [. . .] It is no longer the factory that makes the market: rather, it's the market – with all its volatile preferences, the customer's perennially fluctuating states of mind – that determine the structure of production and production choices."[7] Thus in the development of the Toyota model, old functions dating from the earlier system were gradually reshaped and took on new connotations, which in some cases were of considerable significance. Interesting examples are provided by the purchasing office, the marketing function, research and development, and, most strikingly, the personnel function. In the case of the personnel function, new industrial relations began to be established that impacted on the conception of professional training, so that on-the-job training began to move toward something more clearly approaching genuine educational improvement. There was also a growing awareness that special attention needed to be devoted to essential organizational coordination in this type of industry, and thus to the resulting process of entrepreneurial governance.

One of the focal points of the Toyota system consists precisely in sales and the information systems that have been designed around the sales aspects. "Selling of cars door-to-door is unique to Japan [. . .] A vast majority of cars in Japan are customer ordered [. . .] But if cars are customer ordered, you might ask, how does the factory cope? Here's what happens.

Factory executives try to make an educated guess about the demand for different versions, colors and so forth. On the basis of this forecast, they establish the plant's build schedule, which they also give to the components suppliers so the latter know what to make. The accuracy of these forecasts obviously depends on how frequently the build schedule is revised. This is typically every 10 days in Japan, compared with every month to 6 weeks in the West. Once the orders come in, the assembler adjusts the build schedule to make the specific cars the customer wants.

Because the Japanese practice just-in-time production, doing so is much easier than in the West, which has much less flexible factories and much longer lead times for ordering parts (which sit around as inventory for a long time before they are used)."[8] It should be underlined that "Japanese dealers keep a much closer watch on trends in tastes"[9] and these fluctuations in tastes are communicated in real time to those in charge of planning factory production. This makes it possible to put into practice the comments by Taiichi Ohno in this regard: "The goal [...] is cost reduction. Therefore, an increasing in efficiency must be achieved by a method consistent with this goal. To eliminate overproduction and reduce costs, it is absolutely necessary that the production quantity equal the required number.

Every Toyota plant produces in accordance with actual demand. Car dealers around the country send their orders daily to the main office of the Toyota Automobile Sales Company in Nagoya. These orders are classified by computer as to car type, model, fuel discharge rate, style, transmission, color, and so forth. The resulting data serve as the basis for production requirements at Toyota's production plants."[10] What this meant is that changes could now be easily introduced in order to adjust quality and create products that would truly reflect the customer's desires. Such an approach stood in dramatic contrast with the Fordist systems, which were based purely on the quantity dictated by mass production.

Basically, then, one of the chief features of the Toyota system was its inversion of the decision-making process as compared to more traditional production methods. This can clearly be seen in the observation that "The Japanese assert that Toyota makes products that have already been sold," while industries that still continue with the mass production model "make products it might be possible to sell."[11]

Within the production process itself, the most substantial changes involved elimination of stocks of finished products but also stocks of parts, thanks to the technique known as just in time. This technique is accomplished through the inverse flow of information, a procedure that was turned into an operative method with the name of *kanban*. In addition, these changes had the effect of introducing what amounted to the concept of total quality, by adopting the procedure known as Kaizen, i.e., the search for continuous quality improvement in the production flow, to be achieved by a number of small improvements that enhanced the products and the production system. The combined effects of *kanban* techniques and continuous control over parts led to direct control of quality (comparable to the concept of total quality), by virtue of which a defective part was never allowed to be sent on to the subsequent process.

The other pillar on which the Toyota system was founded consists of the process defined as the "self-activation" of production in reference to operative machines. Taiichi Ohno, the creator of these organizational innovations, offers the following description: "The basis of the Toyota production system is the absolute elimination of waste. The two pillars needed to support the system are just in time and autonomation, or automation with a human touch.

Just in time means that, in a flow process, the right parts needed in assembly reach the assembly line at the time they are needed and only in the amount needed. A company establishing this flow throughout can approach zero inventory."[12] And he specifies, finally, that "With the possibility of acquiring products at the time and in the quantity needed, waste, unevenness and unreasonableness can be eliminated and efficiency improved."[13]

The means by which these techniques were concretely put into practice is that of "*kanban.*" "A *kanban* ('tag') is a tool for managing and assuring just-in-time production [...] Basically, a *kanban* is a simple and direct form of communication always located at the point it is needed. In most cases, a *kanban* is a small piece of paper inserted in a normal even-shaped vinyl envelope. On his piece of paper is written how many of what part to pick up or which parts to assemble.

In the just-in-time method, a later process goes to an earlier process to withdraw needed goods, when and in the quantity needed. The earlier process then produces the quantity withdrawn. In this case, when the later process goes to the earlier process to pick up, they are connected by the withdrawal or transport information, called *withdrawal kanban* and *transport kanban* respectively. [...]

Another role is the *in-process*, or *production ordering kanban*, which tells the operator to produce the quantity withdrawn from the earlier process. These two *kanban* work as one, circulating between the processes within the Toyota Motor Company, between the company and its affiliates, and also between the processes in each affiliate."[14] In this way, every production process in the Toyota context is regulated and governed by *kanban*, an essential method of communication for just-in-time production, which has the effect of eliminating stocks and therefore also the corresponding stockpiling in warehouses.

The integral application of just in time also ushered in a revolution in the purchasing function. One of the typical features of the new approach was a pronounced outsourcing of production according to a rationale that assigned priority to the buy conception – buying inasmuch as possible from outsourced enterprises – as compared to the Tayloristic approach that was based on make, i.e., on making as much as possible inside the factory.

The upshot of the Toyota system was the development of what has been called the macroenterprise, i.e., the system of supplier enterprises which together with the reference enterprise contribute to the creation of the final product. The enterprises belonging to the macroenterprise are no less subject to just-in-time logic than the reference enterprise itself, and this entails a need for the purchasing function to obey specific forms of organizational coordination. However, at first there was no immediate awareness of the need to set this form of organizational coordination in place, as Ohno himself explains: "There was no manual and we could find out what would happen only by trying. [...] In 1963 we started handling the delivery of parts ordered from outside. It took nearly 20 years. Today we frequently hear a chassis maker asking the cooperating firm to bring parts just-in-time as if 'just-in-time' was the most convenient system. However, if used for picking up parts ordered from outside without first changing the production method within the company, *kanban* immediately becomes a dangerous weapon."[15]

The Japanese contracting system has thus become a fundamental element of the entire organizational process. For this reason, the purchasing function and its operative office have come to play an eminently strategic role, which is expressed by means of a global governance of the contracts stipulated with regard to each tender or each purchase of parts. According to Benjamin Coriat, the main results accomplished by this new state of affairs can be summarized in the following words: "The contract is a long-term relationship, whose duration is determined by the life-cycle of the products; it is an institutionalized and hierarchical, contractualized relationship [. . .], governed by special procedures; finally, it is a relationship that favors and internalizes innovation."[16] Coriat himself explains the extent of the phenomenon: "If the argument is considered in terms of added value, it has to be admitted that Toyota outsources roughly 70% of the total value of vehicles sold, while General Motors contracts out no more than 30% of its production."[17] The implications also affect the entrepreneurial organization itself: "[...] the decentralized and outsourced form of the enterprise [. . .] seems better suited to the new forms of competition.

The network enterprise that springs from this change, by outsourcing and contracting substantial dimensions of the enterprise's activity, thus combines the benefit of strong centralization of constraints on change with more rapid adaptability, through decentralization of the initiatives embodied in the response. Unlike the integrated enterprise, whose capacity to respond is more sluggish, the network of contractual relations prompts the formation of a relational rent [. . .]."[18]

It is further worth pointing out that the concept of network enterprise was reached only after a prolonged series of organizational adjustments. Once again, the Japanese experience is instructive, in particular as a background framework within which to examine the account of the Italian automobile industry given by the Italian author Enrico Autieri. Autieri notes that the endeavor to reduce non-indispensable coordination costs prompted operations of outsourcing and tertiarization which eventually led to the concept of the "transactional enterprise." This in turn gave rise to the "network enterprise,"[19] resulting in the transition from what had been defined, in some contexts, as the "short enterprise" – a term which had been adopted to describe the decrease in vertical levels of organization that was a feature of such enterprises – to the current network enterprise.

In the short enterprise there was a tendency, above all as regards the information flow, to replace an extended vertical and hierarchical chain with a horizontal organization, where interactive and transactional relations developed. Therefore, the short enterprise can be described as one that seeks in all aspects to award priority to the concept of flexibility, this concept even being applied to human resources. "[...] our actions should change to suit changing situations. In industry, it is important to enable production people to cope with change and think flexibly."[20]

Here too, the comparison with Western production systems offers interesting food for thought: "[...] Just as [. . .] [the American firm] represented the efficient form of in-house management of mass production of standardized products, the Ohnist enterprise appears as a particularly effective form of in-house management of innovation and change. [. . .] Thus Taylor and Ford are associated with the type of enterprise corresponding to the type of capitalism characteristic of the mass production era, while Ohno belongs to the era of differentiated and quality production."[21] In other words, "Large-scale production of rigorously identical products versus small batches of diversified products."[22]

What should be noted is that "Toyota did not wish to vertically integrate its suppliers into a single, large bureaucracy. Neither did it wish to deintegrate them into completely independent companies with only a market-place relationship. Instead, Toyota spun its in-house supply operations off into quasi-independent first-tier supplier companies in which Toyota retained a fraction of the equity and developed similar relationships with other suppliers who had been completely independent. [. . .]

In addition, Toyota often acts as banker for its supplier group, providing loans to finance the process machinery required for a new product. [. . .]

[. . .] Consequently, the Toyota suppliers were independent companies, with completely separate books. They were real profit centers, rather than the sham profit centers of many vertically integrated mass-production firms. Moreover, Toyota encouraged them to perform considerable work for other assemblers and for firms in other industries, because outside business almost always generated higher profit margins"[23] But this was not all, as the supply relation with other enterprises that were operating on the market at the same time allowed greater circulation and dissemination of innovation. Each of these enterprises was called upon to adapt and be flexible in its quest for innovation, upgrading its own research and development processes not only to the specific requests put forward by Toyota but more generally to the needs expressed by other customers present on the markets. Such needs then acted as the launchpad for further innovations that would later have positive spin-offs for Toyota itself. As Coriat states, "Thus organizational innovation within the enterprise was enhanced by organizational innovation in relations among enterprises, with the result that they mutually reinforced their effectiveness."[24]

These activities provided further justification for the increasingly strategic connotations of the purchasing offices, since relations among enterprises were inextricably linked to questions of organizational coordination and governance of behavior. The widespread circulation of innovation allowed great operative flexibility, but this was always accompanied by special attention to the quality of the products transferred.

In conclusion, production as imagined by Ohno took place in "[...] what we could define as the 'minimal factory,' i.e. a factory where production functions, fixed capital and labor are reduced to the coefficients that are strictly necessary to deal with daily or weekly demand. But it should be kept in mind that in Ohno's vision the minimal factory is above all a factory with 'a minimum labor force' "[25] Essentially, then, it can be stated that "[...] in Ohno stock has the same heuristic value as the 'one best way' in Taylor. For Taylor, starting out from a thousand different ways of performing the same job in order to identify the 'one best way' represents the spearhead of rationalization and the intensification of work. In contrast, the idea of starting out from the stock to identify excess labor is the way Ohno spearheaded his new approach."[26]

The Toyota spirit also features a number of special aspects on the question of price setting: "while in traditional firms price is usually determined by summing costs and the desired profit (that is to say, it is the management that 'decides' the price by starting from costs, considered as an independent and unmodifiable variable), at Toyota it is profit which is determined by subtracting costs from the sale price (in other words, it is the consumer who 'chooses' the price by expressing his or her own preferences, while the profit is a function of the cost compression the enterprise succeeds in achieving)"[27] In the Toyota system the search for profit was thus implemented by pursuing the goal of constant reduction of costs and losses, a goal that was to be attained by means of his philosophy based on the two pillars of just in time and "self-activation" (autonomation). As Ohno himself indicates, "The other pillar of the Toyota production system is called autonomation – not to be confused with simple automation."[28] Ohno provides the following explanation of this phrase: "It is also known as automation with a human touch."[29] "The Toyota production system utilizes autonomation, or automation with a human touch, rather than automation. Autonomation means transferring human intelligence to a machine. [. . .][30] For autonomation to be effective, we must implement a system in which the machines sense the occurrence of an abnormality and stop themselves.[31] Autonomation prevents the production of defective products [. . .], and automatically stops abnormalities on the production line [. . .][32] In other words, [it represents] a device capable of making a judgment [. . .] built into the machine.[33] To produce quality products 100% of the time, innovations must be made to tools and equipment in order to install fool-proofing devices for the prevention of defects. This is called *baka-yoke*.[34] The reference here is to the type of prevention systems that endow machines with something that could almost be likened to a touch of human sensitivity: 'An operator is not needed while the machine is working normally. Only when the machine stops because of an abnormal situation does it get human attention. As a result, one worker can attend several machines, making it possible to reduce the number of operators and increase production efficiency.'[35]

In fact, Ohno goes so far as to claim that the key to his system consists in endowing the machine with human intelligence. In this way, human intelligence, or a human touch, is given to the machines."[36] This also means knowing how to set up the relation that was described in the first chapter of this book as the "assignment of technological responsibility." This assignment ultimately established a close symbiosis between the operator and the machine, the latter having been rendered, as it were, "smart." It is therefore crucial to work toward what Ohno defines as an "information technology" mentality that will supersede the "industrial" mentality just as the latter superseded the "agricultural" mentality in an earlier era. But in this perspective it is equally important to be watchful so that machines do not acquire the upper hand over human beings. As Ohno puts it, "We use the computer freely, as a tool, and try not to be pushed around by it. But we reject the dehumanization caused by computers [...]."[37]

The Toyota system also opened up the possibility of practical application of the concept of a network organization chart within the enterprise. Such a development would subsequently allow extensive flexibility in production programs through meticulous marketing activities – likewise carried out just in time in the sense that they tended to reflect the evolution of customer tastes in real time. The consequences affected both production and organization within the industrial factory, focusing all corporate processes on the aspect of attention to the customer.

This extended development of the Toyota type of organization had its natural breeding ground and nurseries in the manufacturing factories, for it was manufacturing that was most drastically affected through the shake-up wrought by the distributed information technology revolution. And it was not long before significant instances of the new setup were observed, when the first applications began to take effect in the automotive factories of the Japanese Toyota group.

However, the revolutionary systems which culminated in the network enterprise models, with pronounced outsourcing of production and with the constant search for ever-increasing flexibility, are not without critical assessments.

First, there is growing awareness of the risk of what is generally defined as a hollow corporation, i.e., an enterprise that has become devoid of any productive content as the result of unbridled outsourcing of its own original production. Of course, the enterprise is thus far more flexible, but in a sense it is also fragile and susceptible to possible forms of competition even by enterprises belonging to its supply chain, once a number of stages of the production process have been contracted out to these suppliers.

Such risks can be averted in several ways, in particular by enhancing the internal research processes through continuous development of profound innovations that will constantly drive the production cycle. Moreover, such innovations should in turn act as an engine of innovation in the organizational conception of processes. Today this can be facilitated by the massive presence of distributed information technology, whereby modifications can be introduced into software programs in response to new interpretations of the best innovative utilization of the hardware machines. In this way, continuous and widespread innovation can be fostered and enhanced precisely by the presence of a strong tiered supply chain composed of a web of suppliers and sub-suppliers.

Thus the personalized product also embodies a wealth of knowledge, information, and know-how, and this sparks further organizational innovations. For example, it increasingly prompts enterprises to introduce concepts of maintenance the customer must be alerted to, in a perspective where the customer is seen more and more as a subject who needs to be educated and instructed in the best way of utilizing and preserving the product. Consequently, a new perspective on knowledge is taking shape within the network enterprise, with knowledge now being reconsidered as a factor that should be included within strategic priorities.

Corporate culture thus needs to become the culture of the individuals who make up the enterprise. Such a conception represents an effective and real antidote to another potential hazard present in the network enterprise, namely the risk of movement away from a unified entrepreneurial conception. Characterized as it is by a network of independent technologies guided by a network of individuals who are likewise to a large extent independent, in a context where the concept of software prevails over that of hardware, the network enterprise may sow the seeds of dangerous centrifugal forces, tending to magnify the distance between the various elements of the supply chain. It is here that culture, interpreted as a strategic project of the enterprise, can mitigate these dangers by recomposing the different forms of entrepreneurial behavior in a unified conception. Thus awareness of the importance of the culture of individuals heralds rediscovery of the human factor as a strategic element, to be considered as a priority element in the entrepreneurial situation. This entails increasing realization of the need for a figure who has to learn to operate as a "self-entrepreneur," i.e., a figure endowed with considerable decision-making and operative independence, yet who seeks at the same time to improve the operative corporate environment, above all in terms of production of widespread innovation.

Culture can thus become the great new frontier of the network enterprise. It is a vision of culture that can also contribute to shaping a trailblazing new strategy of governance and cohesion of the enterprise. This explains why, when addressing the issue of the network enterprise, it is crucial to revisit the strategies and examine the way they are formalized as specific strategic plans in order to address the challenges of the new entrepreneurial dimensions.

Notes

1. Benjamin Coriat, *Ripensare L'organizzazione del lavoro. Concetti e prassi nel modello giapponese* [*Rethinking the Organization of Labor. Concepts and Practice in the Japanese Model*] (Italian translation of Benjamin Coriat, *Penser á l'envers. Travail et organisation dans l'entreprise japonaise*, Paris, Christian Bourgois, 1991), Dedalo, Bari, 1991, p. 54.
2. Taiichi Ohno, English translation *Toyota Production System*, Productivity Press, New York, 1998, p. 1.
3. Guido Rossi, *Il Gioco delle Regole* [*The Rules Game*], Adelphi, Milan, 2005, p. 20.
4. Taiichi Ohno, *Toyota Production System*, cit., p. 15.
5. Ibidem.
6. Ibidem, p. 30.
7. Marco Revelli, *Introduzione* a Taiichi Ohno, *Lo spirito Toyota* [*Introduction* to Taiichi Ohno, *The Toyota Spirit*] Einaudi, Turin, 1993, xxix.
8. James P. Womack, Daniel T. Jones, Daniel Roos, *The Machine That Changed the World*, Harper Perennial, New York, 1991, p. 182.
9. Ibidem, p. 183.
10. Taiichi Ohno, *Toyota Production System*, cit., p. 62.
11. Marco Revelli, *Introduzione* a *Lo Spirito Toyota* [*Introduction* to Taiichi Ohno, *The Toyota Spirit*], cit., xxx.
12. Taiichi Ohno, *Toyota Production System*, cit., p. 4.
13. Ibidem, p. 123.
14. Ibidem, p. 124.
15. Ibidem, p. 32.
16. Benjamin Coriat, *Ripensare l'organizzazione*...[*Rethinking the Organization*...], cit., p. 119.
17. Ibidem, p. 122.
18. Ibidem, p. 139.
19. Enrico Auteri, *Dalla Gerarchia alla Leadership* [*From Hierarchy to Leadership*], Guerini, Milan, 2004, p. 301.
20. Taiichi Ohno, *Toyota Production System*, cit., p. 52.
21. Benjamin Coriat, *Ripensare l'organizzazione*...[*Rethinking the Organization*...], cit., p. 169.
22. Ibidem, p. 22.
23. Womack et al., *The Machine That Changed the World*, cit., pp. 61–62.
24. Benjamin Coriat, *Ripensare l'organizzazione*...[*Rethinking the Organization*...], cit., p. 136.

25. Ibidem.
26. Ibidem, p. 25.
27. Marco Revelli, *Introduzione* a *Lo Spirito Toyota* [*Introduction* to Taiichi Ohno, *The Toyota Spirit*] cit., xxx.
28. Taiichi Ohno, *Toyota Production System*, cit., p. 6.
29. Ibidem.
30. Ibidem p. 121.
31. Ibidem, p. 113.
32. Ibidem, p. 122.
33. Ibidem, p. 121.
34. Ibidem, p. 122.
35. Ibidem, p. 7.
36. Ibidem, p. 6.
37. Ibidem, p. 48.

Chapter 7
Strategies for the Network Enterprise

In the light of the risk of a hollow corporation, as mentioned in the previous chapter and given above all the tendency toward excessive decentralization of operative responsibilities – with the ensuing dissolution of unitary entrepreneurial action – it becomes vital to focus attention on the significance of strategies. Strategies become the key element for adequate governance of corporate behavior.

Moreover, special importance attaches to the strategic guidelines and the mid- or long-term plans drawn up on the basis of these guidelines. Only by clear delineation of the strategic guidelines and adherence to a pre-established plan does it become possible to assure proper management of all the entrepreneurial activities.

The concept of "strategy" can be traced back to ancient origins, originally deriving from military culture. For instance, strategy is mentioned on numerous occasions by ancient Chinese philosophers, and it was taken up again at the beginning of the 19th century by General Karl Von Clausewitz, who "defined strategy as an activity that makes it possible to regulate and coordinate the military activities in order to achieve the aims of the war; strategy is thus distinct from tactics, the latter being the art of arranging and directing the battle as an independent unit. Therefore, military strategy determines the objectives to be assigned to the armed forces so that the final goal is achieved with the greatest success in the least time and with the least sacrifice."[1]

The concept of strategy was subsequently applied to the entrepreneurial framework, where it was seen as an expression of the best way to act. A strategy, it was argued, was the appropriate manner to balance pursuit of the objectives of the enterprise against the constraint of available resources and the general environmental conditions in which the enterprise has to operate. This approach, in turn, brought the concept of an objective itself into sharper focus and led to clearer awareness that implementation of a strategy to achieve the objective would in itself be influential in shaping the organizational structure of the enterprise.

This situation was cogently portrayed by Alfred D. Chandler, who in his fundamental 1962 work – *Strategy and Structure* – pointed to a substantial affinity between strategy, on the one hand, and planning and management of entrepreneurial development on the other hand. In Chandler's definition, strategy is seen as "the determination of the basic long-term goals and objectives of an enterprise, and the adoption of courses of action and the allocation of resources necessary for carrying

G. Dioguardi, *Network Enterprises*, Innovation, Technology, and Knowledge Management 104, DOI 10.1007/978-1-4419-1333-3_7, © Gianfranco Dioguardi, 2010

out these goals."[2] The significant conclusion reached by the Bostonian scholar is that "structure follows strategy and that the most complex type of structure is the result of the concatenation of several basic strategies."[3]

It goes without saying that there are certain fundamental aims that any type of enterprise must always pursue. These include survival, growth, and therefore, necessarily, profit. More specifically, strategy must be able to respond to a plethora of problems such as what direction to take (and thus also a possible modification of the market segments targeted by the entrepreneurial action), what to undertake in terms of quantity and quality of production, how to engage in product diversification, or the most suitable level of price setting in the case of products to be launched on the market. In addition, strategy must be able to identify and orientate various aspects of corporate behavior within the enterprise itself.

Naturally, particular attention must be devoted to the macroenvironment and its sub-systems, namely its economic, socio-demographic, sociocultural, political, and legislative system, the basic technology characterizing the macroenvironment, the national structure, and the Country system.[4] But it is also essential to analyze the state of the enterprise in its organizational and structural components. Here it may be useful to refer to the model proposed by Larry Greiner, which sees the history of the enterprise as set between periods of evolution (i.e., prolonged periods of growth, during which the organizational settings are not thrown into turmoil) and periods of revolution (in which various disruptive events take place in corporate life). Greiner classifies these periods as consisting of five developmental stages linked mainly to phenomena of creativity, authority, assignment of responsibilities, coordination, cooperation; each stage terminates with a corresponding moment of crisis (pertaining to command, independence, control, bureaucracy, and a final crisis that the author leaves undefined).[5]

The ideas springing from analysis of the ongoing strategies have to be incorporated in a rational manner in the mid- and long-term strategic plans. This is necessary in order to shape the guidelines of corporate behavior which, especially in the network enterprise, must closely involve each of the participants who contribute to the entrepreneurial event.

A strategic plan that looks beyond the short term, generally a medium-term plan, is a tool that allows all the figures taking part in the network enterprise to determine what type of behavior (strategies) is likely to be the most appropriate to achieve the proposed objectives, taking into account the circumstances involved in the economic scenario. Therefore, the strategic plan can also be seen as a network of support decisions that back up the operative development plan.

During the present-day period, strategic plans generally take shape in an environmental scenario that is characterized by complexity and turbulence. As a result, continuous adjustments are required, not only in the objectives but also in the actions to be undertaken in order to achieve them. This entails that every corporate planning tool must have pronounced operative flexibility, constantly allowing for innovation in the expected forms of behavior.

With regard to identification of the objectives to be pursued, a priority role is generally assigned to the demands voiced by the stakeholders, that is to say, by all those who have some form of vested interest in the network enterprise. Therefore, careful attention needs to be paid to exploring the actual scale of priorities inherent in the real and potential demands manifested by the stakeholders themselves, so that all the players who are protagonists of the strategic plan can feel that they truly share in the overall corporate aims. Concretely speaking, then, there must be a preliminary endeavor to acquire information from the grass roots, which will then be processed by those in charge of the strategic plan. This is implemented with a procedure reminiscent of the Delphi method, a bottom-up method, which then restores to each operator, in a top-down sense, the information that has been processed and the decisions made.

Utilization of the plan requires constant monitoring of the outcomes and of the targets achieved. Such an approach entails monitoring and control during the short term, by means of strategic planning that converts medium-term targets into short-term stages. This is essential in order to carry out an in-depth step-by-step analysis of any disparity between the planned objectives and the results actually achieved. Only by careful stepwise monitoring does it become possible to promptly undertake corrective retroactive control measures (feedback) that adjust either subsequent actions or, if necessary, even the targets themselves – a measure that may well become necessary in the case of severe disruptions in the background scenario. The aim is to conclude the entire planned period with a balance sheet that can be considered satisfactory in the light of the results effectively achieved. So efforts will be made to "construct" effective planned control, projected into a future that is actualized little by little according to the requirements expressed by the plan.

To ensure satisfactory progress of the strategic plan, it is crucial for all the management functions and all the operators to be actively and constantly involved in the corporate activities. This can be accomplished by proper use of the planning procedure together with bottom-up–top-down processes, as well as appropriate information transfer and circulation of details on decisions taken.

Overall, mid- and long-term strategic plans have to be based on a preliminary analysis of the general scenario within which the plans themselves are to be implemented. However, in this era of distributed information technology, the typical scenario is one of continuous and rapid variability, making it difficult to provide an exact static representation of a constantly evolving picture. Therefore, it is highly important to try to operate through forward-looking visions that seek to anticipate the future.

In a very general perspective it can be argued that the elements characterizing these scenarios have a twofold nature. Some of them can be considered as being within the sphere of power of the network enterprise that acts in the given scenario. In such a case, the elements of the scenario can be influenced by the behavior laid down in the strategic plans, while the plans themselves can to some extent anticipate the future through governance of the performances achieved by the organisms in charge of the plans. By contrast, there are other elements that are beyond the power of entrepreneurial influence, in which case it is these external elements that necessarily affect what kind of forecasts are predicted. Yet the enterprise cannot shy from a proactive response – which it must necessarily engender in order to avoid being a prisoner of purely passive reactions – and this it must do by exploiting the flexibility built into the strategic plans precisely to counter such adverse situations.

Another important aspect concerns specific analyses of the behavior of the potential competitor enterprises (benchmarking), in order to highlight the social, economic, technological, and political forces that can exert competition and are likely to have a positive or negative impact on the environmental factors. These analyses should make it possible to define the strengths and weaknesses of the network enterprise, so as to aid the introduction of continuous performance improvements.

From an environmental point of view, one of the most important aspects is that of identifying the main strategic events or the tendencies of emergent scenarios that could lead to risks and/or uncertainty in the framework in which the network enterprise is operating. Risks and uncertainty spring above all from the economic framework, but they may also relate to the environmental components – understood here as the sociocultural and political legislative setting – that are present at the local level. However, the analysis will also take into consideration the state of technology and the degree of innovation present.

The forms of behavior of the network enterprise are certainly influenced by the stakeholders, represented above all in particular by opinion groups, banks and credit institutions, trade unions and workers in general, but also state bodies and parts of the state administration. Within the enterprise itself, further groups that have to be taken into account are the operative staff and the corresponding organizational structure, especially as regards central staff and top management.

By monitoring these interactional phenomena, more concrete predictive descriptions of strategic processes can be drawn up, through better definition of the evolutionary scenarios. This will highlight some of the more influential factors, with their related predictable environmental risks, while allowing a margin of evaluation for unpredictable risks. It should also become possible to prefigure potential trends springing from past situations that may conceivably be destined to burgeon in the future.

Although special attention has to be paid to starting conditions, one of the key elements is definition of the mission of the network enterprise: in fact, defining its mission constitutes the first step toward working out its strategies. In this perspective, it is necessary first and foremost to establish the most specific ends to be pursued and then to proceed toward identification of the objectives.

Together with the mission, another major aspect consists of the vision the enterprise wishes to embrace. Among the many definitions that have been put forward to give an effective portrayal of corporate goals, the following offers a particularly apt statement: "The mission is a wide-ranging statement of the goals the enterprise is pursuing. It generally identifies broad areas of activity in the economic and social field."[6] And as indicated more specifically by Lucio Sicca, "The mission is, strictly speaking, a statement of the current situation and of future projections concerning product extension (supply), market extension (segment of demand), geographic extension (markets where the product is present) and the exclusive know-how developed (and to be developed) by the enterprise in order to achieve an advantage that can be defended in the long term."[7] But the concept of a mission also means defining the image the enterprise hopes to communicate as it moves toward the future. This indeed is its vision, which is once again fittingly expressed by Sicca: "[...] the vision is the statement of the prospect the enterprise is working towards; it is what the enterprise could/would like to become, an idea of the expected future. [...] [V]ision – the prospect of 'what' the enterprise will be able to become, the image of a desired future – encompasses the concept of 'mission', which is an expression of why an enterprise exists and what role it should play in civil society (mission in the broad sense) and the extension (scope) of its presence (mission in the narrow sense)."[8]

Basically, then, mission and vision define "where the enterprise is going and what identity it wishes to assume."[9] For example, the Ford Motor Company sets itself the mission of "introducing continuous improvements in our products and services in order to satisfy the needs of our customers, thus allowing us to prosper as a firm and to provide a reasonable return to our shareholders, the owners of our company."[10] Immediately thereafter comes a definition of the values the company seeks to adhere to and the guiding principles that should provide the fundamental inspiration for the entrepreneurial action. An interesting example of the concepts that inspire a company's mission is found in a statement by the Pirelli Company (Pirelli S.p.A.), which sets itself the aim of "[...] producing and marketing complex high quality products and operating successfully from our various international bases. We assign great priority to continuous improvement of the environmental impact of our processes and products."[11]

What is clear from the above description of mission and vision is that these two concepts delineate possible strategies of differentiation, of pursuit of quality and internationalization of markets. Some authors such as Giorgio Pellicelli identify mission as consisting in the "definition of four elements: the main goals the enterprise intends to achieve; the area or areas of business in which the enterprise intends to compete; the policies and values that inspire the enterprise; the behavioral rules to be followed in decisions involving social responsibility."[12] And in order for such an approach to be meaningful, there must be clear awareness, emerging from these situations, of the values the members of the personnel should believe in and the behavioral rules that should provide the inspirational background for their entrepreneurial action.

These, then, are the main guiding elements of the strategic plan, which must thus be established as a function of the objectives and goals the enterprise wishes to achieve.

Before going into the specific objectives in detail, it is worth dwelling for a moment on some aspects of a general nature that should provide the underlying buttress and inspiration for the objectives themselves and the ensuing operative behavior.

The basic objective of the network enterprise, understood as an innovative structure, must be that of presenting itself as a "center of excellence," in order to counter the generalized complexity observed in the environmental scenario and the world of enterprises. In this sense it is vital for there to be a constant quest for a "high" road leading to development of individuals' actions and facilitating the pursuit of success, especially as a means to guarantee the satisfaction of all the stakeholders. This implies that the network enterprise must pursue constant learning in the field. One of the main ways in which its strategic management can be implemented is by operating analogously to a learning organization that is capable of increasing its own knowledge and competencies, thereby promoting the cultural growth of its staff. A distinctive feature of the network enterprise will thus be the production and sharing of knowledge among all the figures taking part in the management phenomenon. The enterprise will also develop a form of culture constantly orientated toward the "outcome," and all the operators present within the network will share a sense of responsibility in pursuing this end.

The teachings of Joseph A. Schumpeter are germane to this point. The enterprise should award priority to goals that promote the introduction of innovations of all kinds (product, production process and organizational innovation, new trends in marketing, training, information and communications, etc.); in addition, the enterprise should seek to set in place processes of ever-increasing internationalization so as to address new segments of the global market, in particular in areas the enterprise has only recently entered. Today this conception is of particular interest, for example, with reference to the emergent markets of China and India, although the consolidated American markets and the variegated European context should by no means be overlooked).[13] Therefore, it is right to support and promote processes of innovation and modernization of the entrepreneurial system through integrated actions of technical assistance, training, information transfer, and consultancy, while also developing industrial research activity as well as innovation and technology transfer.

The enterprise will reinforce its systemic interdependencies by using advanced technology (ICT) and in particular by adopting appropriate telematic networks designed to integrate the enterprise's own operative network, in order to provide increasingly effective and efficient services. This in turn will make it possible to pursue a conception of the network enterprise as an integrated service system, where there is a constant interchange of experiences between the center and the periphery.

Let us now look more closely at the objectives of the enterprise. Its major objectives can be subdivided into several macroareas: those linked to operative management of an internal type (structure and organizational strategy) and those of an external type specifically linked to the market and therefore also to the survival of the enterprise.

Achievement of the objectives necessarily brings about a change in the scenario where the enterprise is operating. And it is precisely this change that in some respects becomes the yardstick for measurement of the efficiency and effectiveness of the efforts made so far. On the other hand, change should also be interpreted as part of a strategy which includes awareness and conservation of tradition, so that the network enterprise is poised in a situation constantly hovering between continuity and novelty. In this framework, what is needed is a twofold emphasis: focusing on the traditional situations that are typical of the enterprise's action, yet highlighting the various lines of innovation. For only through constant change and renewal can tradition be preserved and revitalized.

The internal objectives should be defined in relation to the organizational structure in such a way as to render the network structure increasingly efficient, with each member of the staff constantly oriented toward the outcome. This underlines the need to place emphasis on the human capital, thereby ensuring that it becomes one of the strengths of the entrepreneurial action. In addition, the highly important aspect of enriching the enterprise by systematic introduction of new typologies of professional skills should not be ignored, nor should it be overlooked that the entire body of personnel must be encouraged to feel a sense of individual responsibility and profound involvement in achievement of the objectives. Therefore, it is vital that all the members of the enterprise should work toward the overall goals by adopting a results-oriented approach. And precisely in order to ensure that this is indeed the predominant attitude among the personnel, with each member working in a context that is most congenial to his or her inclinations, it is very helpful to decentralize the decision-making process with a view to enhancing the development of the network structure. To this end, the general aim should be to rationalize the organizational structure and, correspondingly, to optimize function attribution to the members of the personnel, thereby also pursuing the further aim of containing costs through an improvement in corporate efficiency. Naturally, this approach should be constantly monitored with a system of management control oriented above all to analysis of costs. At the same time, however, there should be a constant endeavor to promote forms of research that generate institutional, technological, process, and organizational innovation to ensure vitality and renewal within the overarching entrepreneurial effort.

Another important consideration relates to the amount and nature of resources the enterprise can genuinely count on. The fundamental resources must include the capital of human resources, with special attention to professional capabilities and competencies. This capital represents the most significant resource possessed by the enterprise. Its store of knowledge and information can embody numerous aspects of knowledge management, with an extremely valuable and absolutely unique capacity of governance that can be crucial for whatever activity the enterprise seeks to undertake. Indeed, it is these typical characteristics of the human factor, and the associated immense store of professional capabilities, that enable the network enterprise to express an efficient organizational setup, which deploys these capabilities in the form of a learning organization that can positively affect the strategic orientation adopted by the management.

Careful attention must likewise be paid to examining the financial and material assets (real estate, machinery, high-tech equipment, etc.), as well as the technological factors, frequently characterized by advanced ICT. Equally important, if not even more so, are the immaterial resources concerning the enterprise's organizational knowledge, which contribute to raising the quality of performance within the management. In this regard, another major aspect concerns the need to devise a motivational system that will encourage all the participants to strive toward an ever-increasing sense of responsibility in the quest for improved quality. On-the-job training, to raise awareness of entrepreneurial culture, will be particularly useful in this respect.

All these aspects, as well as the individual history of each network enterprise – a history that has been built up by accumulated experience, with the present standing on the shoulders of giants of the past – contribute to increasing the operative potential and thus also the competitiveness of the enterprise. For this reason, even competitive capacity should be considered as another strategic resource, to be exploited as effectively as possible.

Furthermore, the information flow system within the enterprise plays a crucial role, especially within the network model. The actual as well as the potential performance of the information flow system should always be carefully examined. The same holds true for the cost control system, which needs to be able to respond promptly and rationally to the increased management requirements of the network enterprise. By appropriate consideration of these aspects it also becomes possible to improve implementation of control over the yearly budgets, leading to greater efficiency in obtaining information, both in quantitative and temporal terms, on which to base the corresponding management decisions.

In the structural sphere, if the network organization is increasingly to become a strategic resource, then it will be necessary to carry out an analysis of its strengths and weaknesses, in order to identify possible threats and potential opportunities. This calls for a clear definition of the various aspects of the core business the network enterprise aims to focus on, with particular emphasis on the services to be offered as a means for obtaining the best competitive advantage on the market.

In terms of internal goals pertaining to the organizational strategies, it may be necessary to reexamine the diversification of the various areas of intervention. It could prove necessary to expand such areas to address new situations that have come to be considered strategic, whereas other areas may have to be conflated, or in fact eliminated, if they are considered too similar to one another or are of no particular interest. Such moves imply the need to conduct appropriate market segmentations in order to implement effective diversification strategies. Naturally, the approach will always be oriented toward achievement of the goal of corporate growth. Consequently, emphasis will be placed on external objectives, which are concerned with those markets where the enterprise aims to build up a specific competitive advantage. In so doing, the enterprise will be able to successfully face the challenges of ever-present competition.

Outlined in the manner suggested above, the objectives are then brought into sharper focus in the detailed strategic plan, where they will be more adequately defined for the short, medium, and long term.

As far as the structural objectives are concerned, the functional organization chart should draw its inspiration from the organizational model of the "network structure," which should form the basis for the organizational design of the enterprise. The model must be able to assure pronounced operative flexibility, with the accumulation of specific competences in each of the nodes, such competencies having been acquired both from the existing corporate culture and also from learning by doing. The latter is to be understood as the body of actions that are performed as part of the daily activities and which effectively transform the network enterprise into a learning organization.

The model must be expressed by a functional hierarchical structure described in the corresponding organization chart, in which the chain of command should be as short as possible (the so-called short enterprise) in order to render the organizational context more efficient. But at the same time, the assignment of responsibilities must be set up as a function of control over the results to be achieved, rather than as a function of the manner of achieving such results. This ensures that the operativeness of the organizational structure will bring about control "by exceptions," i.e., control over data that diverge from expected results. The specific activities will be left to the operative innovation of individual members of the personnel, who will in fact have to operate as if they were "self-entrepreneurs," and also as knowledge workers, working on the basis of constantly acquired competencies. Moreover, it is vital to ensure that decentralization and the associated centralized coordination always develop through forms of participative involvement which promote cohesion and integration, so as to create a bond that links the various subjects to the entrepreneurial governance network. In this way the functional hierarchy sketched in the organization chart will effectively operate through the individual transactions and interactions, leading to forms definable as a "quasi-market." Here the workers will each directly decide on the mode of action best suited to performing the task assigned to them, operating with the full-blown independence dictated by the decentralization typical of the network organization.

At the same time, however, the network of individuals must be guided by a constant governance, so that each node of the network is accountable for the results achieved. This implies that each node must shoulder responsibility for harmonizing the pursuit of efficiency and effectiveness with achievement of the overall corporate aims.

If the network enterprise is interpreted as an integrated system, then the strategy expressed in the strategic plan also has to allow for its own adaptability to environmental change. Consequently, it is quite natural to think that construction of the plan is something in a constant flow of becoming, driven by reflection on action that is an intrinsic part of the day-to-day activity of all the components of the system, including the external stakeholders.

This means that the strategic plan must assume a dynamic character, in constant transformation. It cannot be defined once and for all. Instead, it should be seen as a factor of transformation that has the power to kindle the dynamics of constant change. And as is typically the case for strategies, here too every moment of realization throws new light on what had been planned, new insight, thereby modifying its meaning and, accordingly, enhancing awareness of the planned course of action and fine-tuning the subsequent steps.

In sum, the plan will have to be realized through a flexible planning model that can be continually updated to take into account the manifold requirements emerging over the course of time. The strategic goals will make it possible to delineate activities for the period under examination and thence to establish planning lines that will act as a *trait d'union* between the strategic context and the actual short-term operative situation. Short-term planning can be developed as a calendar-based agenda of the main events (event budget), which will then be completed by the specific quantitative and more strictly administrative budgets pertaining to assets and liabilities and the economic balance sheet.

Strategic planning in the network enterprise, then, is shaped around the definition of actions aimed at achieving the goals enshrined in the enterprise's mission and vision. The strategic objectives should be planned at yearly intervals in order to ensure that they can be fully accomplished during the period of time examined by the strategic plan.

Special attention should be devoted to the possibility of designing a service that endows the enterprise with an ethical and cultural dimension. This is of immense importance as a means of enhancing in-house professional capabilities, which in turn act as a driving force for consensus and motivation among the workers. Therefore, one might suggest the introduction of a purpose-designed "Culture Project" to initiate a cultural consciousness raising and corporate ethics campaign. This could be done, for instance, by arranging periodic cycles of lectures to be held at intervals throughout the period of time considered. Such a proposal would be in line with the need for the network enterprise to present itself in its day-to-day management as a socially concerned enterprise, and on this issue there should always be exemplary transparency. It would be useful, in this connection, especially on the plane of vision, to set up a special "Social Budget" which would, in a sense, be complementary to the traditional assets and liabilities statement and financial budgets. The Social Budget would be entrusted with the task of representing all the demands voiced on such matters: it would be a tool specifically set up to address the variegated issues that emerge along the evolutionary path set in motion by the strategic plan. It would thus constitute a further tangible and important resource, of significance both for the in-house and external relations of the network enterprise. So even this rather special budget could play its part alongside the traditional economic and financial budgets by underlining the social activities and ethical commitment of the enterprise during the management period considered. Through the Culture Project, with the associated Social Budget, it becomes feasible to pursue the important objective of concretely orienting the enterprise toward the desirable concept of a "socially concerned enterprise."

Notes

1. Lucio Sicca, *La gestione strategica dell'impresa. Concetti e strumenti* [*Strategic Management of the Enterprise: Concepts and Tools*], Cedam, Padova, 1998, p. 55.
2. Alfred D. Chandler, Jr., *Strategy and Structure: Chapters in the History of the Industrial Enterprise*, MIT Press, Cambridge, MA, 1962, p. 13.
3. Ibidem, p. 14.
4. Lucio Sicca, *La gestione strategica dell'impresa...*[*Strategic Management...*], cit., p. 348.
5. Greiner, *Evoluzione* [*Evolution...*], cit. Cf. also the work by Gianfranco Visconti, "Il modello di Greiner applicato a un'azienda in espansione" ["Greiner's Model Applied to an Expanding Company"], in *Bari Economica* n° 3 (2005), pp. 55, 65.

6. Giorgio Pellicelli, *Management, 3 Strategia* [*Management, 3 Strategy*], Joint edition by *La Repubblica* and Università Bocconi, Il Sole 24 Ore, Milan, 2005, p. 54.
7. Lucio Sicca, *La gestione strategica dell'impresa...*[*Strategic Management...*], cit., p. 722.
8. Ibidem.
9. Giorgio Pellicelli. *Management...*, cit., p. 56.
10. In Lucio Sicca, *La gestione strategica dell'impresa...*[*Strategic Management...*], cit., p. 89.
11. Ibidem, p. 90.
12. Giorgio Pellicelli, *Management...*, cit., p. 57.
13. Federico Rampini, *L'Impero di Cindia* [*The Empire of Cindia*], Mondadori, Milan, 2006.

Chapter 8
Ethics and Social Values in the Network Enterprise

The concept of the socially concerned enterprise is gaining increasing credence at this time, and its range of connotations is expanding to the point that it can almost be likened to an action of entrepreneurial marketing. Moreover, the new meanings assumed by the behavior of the individual in the network enterprise redefine the values to be considered in entrepreneurial action and thus even contribute to redefining the very concept of corporate ethics. Indeed, if the enterprise is viewed as a set of individuals who are strongly motivated toward achievement of common goals, then the overall ethical vision expressed by the enterprise can play a significant role by seeking to promote morally motivating values. This will not only encourage positive co-existence among persons who have to work together, but it will also favor increased operative effectiveness and efficiency.

As was noted earlier, one vital aspect of network enterprise corporate ethics is that the members of the enterprise must be prompted to interpret the role of self-entrepreneurs, underpinned by exactly the same ethical motivations that characterize the entrepreneurial leader. Once measures to encourage this attitude are in place, they will also bring about an atmosphere of constant cognitive learning, with the result that the steadily growing body of knowledge accumulated by the individual workers will gradually be pooled. A satisfying and motivating corporate culture with appropriate behavioral values will be built up.

This new form of ethically based corporate culture can also help to improve the external context. It should never be forgotten that the enterprise is rooted in its surrounding environment and should pursue the aim of constantly improving this environment. Culture and ethics are thus destined to be transferred toward the outside not only through the action of individual workers but also through the general behavior of a network enterprise that is more and more aware of its social context and the meaning of corporate ethics.

The increasingly widespread acceptance of the concept of "social responsibility of the enterprise" is accompanied by a growing tendency to view it not only as an innovative cultural approach but also as a mark of quality. For instance, commenting on the new "*enterprise citizenship*," Stefano Salis states among other things that "The social commitment of enterprises is the new phenomenon that entrepreneurs, associations and above all normal citizens will increasingly have to be concerned with. [. . .] Now even small and medium-sized enterprises – the real connective

G. Dioguardi, *Network Enterprises*, Innovation, Technology, and Knowledge Management 104, DOI 10.1007/978-1-4419-1333-3_8, © Gianfranco Dioguardi, 2010

fabric of Italian industry – are more and more determined to insist on the capacity to have an impact on the social side [...]. He then also notes that 'socially responsible enterprises are more dynamic and benefit from greater appreciation by their workers, customers and investors. In a word, more competitive [...].' "[1]

This leads to the rediscovery of ethics as a phenomenon that belongs both to the individual and to the "enterprise subject," the latter being taken to refer to second-level individuality. Indeed, the enterprise – and in particular the socially concerned enterprise – is itself a great storehouse of values that characterize and guide the enterprise in its strategic choices.

It is of interest, in this regard, to devote a moment's attention to an analysis of the meanings of the terms "value" and "values" insofar as they are used with specific reference to the enterprise. These are by no means synonymous terms, yet a number of often hidden linkages can be brought to light.

In an increasingly turbulent world that bristles with fiercely competitive firms, the system of social values must become one of the key elements of entrepreneurial success. It is imperative that this system should be capable of infusing fresh lifeblood into the enterprise spirit, thereby also revitalizing the value the enterprise constantly creates. But it is not merely a case of revitalization: survival itself is at stake, and a proper system of social values can become a driving force for strategic survival as well as for correct professional ethics, as is made clear by a glance at the history of the enterprise in its thrust to move forward into the present-day context, where the dominant features are the IT revolution and the high-tech world. In particular, within the entrepreneurial sphere it is imperative for values to be interpreted as resources capable of constituting competencies in the new environmental scenario, characterized as it is by knowledge development. This is all the more true when the scenario in question has a network conformation, whether one is dealing with large or small enterprises: in either case, survival is closely linked to the spirit that characterizes the enterprise and the values that shape its development.

One may well be surprised at the range of meanings that attach to the word "value," even when it is declined in the plural.

In economics, "value" immediately evokes the meaning of price or cost, and it calls to mind the most classical of definitions, that of Adam Smith (1723–1790) in his work *An Inquiry into the Nature and Causes of the Wealth of Nations:* "The word value, it is to be observed, has two different meanings, and sometimes expresses the utility of some particular object, and sometimes the power of purchasing other goods which the possession of that object conveys. The one may be called 'value in use'; the other, 'value in exchange.' The things which have the greatest value in use have frequently little or no value in exchange; and, on the contrary, those which have the greatest value in exchange have frequently little or no value in use."[2] He then proposes the example of water, which is rich in "value in use" but has only scanty exchange value, and also the case of diamonds which, although of little practical utility, are important means of exchange.

Very many authors have studied these issues in great depth and have put forward other definitions, from the classical theories of Ricardo (1772–1823) and

Malthus (1766–1834), to Marx (1818–1883), who reexamined the characteristics of "value" in the light of the "quantity of labor contained." Later one also finds the clarifications of Jean Baptiste Say (1767–1832) and John Stuart Mill (1806–1873), and subsequently the subjective theory of value and the commentaries by Marshall (1842–1934), Pareto (1848–1923), and numerous others. Claudio Napoleoni summarizes the question as it came to be delineated over time: "The content of the concept of value has been far from constant over the history of economic doctrines. On the contrary, on occasion the change affecting this content has been so far-reaching that a list of the definitions adopted at the various different times might leave the reader faced with genuinely different opinions [...]."[3]

With reference to "values," in particular those of entrepreneurial organizations, the meaning may be as defined in the *Grande Dizionario della Lingua Italiana*: "Anything which is worthy of appreciation in the ethical sphere, whether in the subjective sense as the object of a preference or of a moral choice, changeable depending on the individuals and the historical eras, or in the objective sense, as a universal principle that is completely independent of relations with man, and which is absolute and pertinent to the sphere of what must be." Accordingly, "in everyday language: everything that is recognized as being worthy of appreciation and thus desirable in an ethical perspective [...]."[4]

Therefore, "values" make reference to the ethical modes of behavior that a subject – an individual or an enterprise considered as second-level individuality – manifests through the operative action adopted.

Both meanings of "value" and "values" are permeated with life through the historical dimension; they are immersed in the history of the subject to which they refer, the latter being, in turn, destined to create constantly new forms of value. However, despite the contradictions that may emerge between the two meanings – economic and ethical – these interpretations are linked by often unnoticed bonds, or "hidden connections," to use the terminology dear to Fritjof Capra.[5]

One might even venture to state something approximating a law according to which "value" is necessarily generated by the "values" the enterprise is founded on. Moreover, since both these elements are infused with life through the history of the enterprise, they depend on the experience the enterprise has built up – a history manifested in a life cycle which, by its very definition, cannot be indifferent to the passage of time. Consequently, it is also affected by a natural obsolescence that consumes its contents, resulting in the need for continuous maintenance in order to constantly regenerate its potential.

This illuminates why the organization of values is manifested as a strategy in order to update the value produced, and the values themselves appear in the guise of entrepreneurial resources and indispensable competences enabling the enterprise to be successful in facing competitive challenges on the markets. Essentially, in the present-day turbulent world, the ethical values that can be seen as the typical elements of corporate behavior must become attributes of success and, therefore, elements of entrepreneurial survival as well as elements of strategic behavior that is correct from the point of view of professional ethics. This approach, which holds true for large enterprises, must also be valid for small enterprises, whose survival, in the scenario of constant change that is so fraught with difficulties, is crucially linked to the spirit these enterprises are able to build up.

No enterprise can afford to ignore that the environment in which the economy operates today is influenced by more and more abrupt and far-reaching changes. The present era is one of great demographic, technological, social, economic, and geopolitical transformations. We are living in difficult scenarios, especially since the time when the great high-tech revolution and the world of IT and communications burst on the scene.

As early as the 1980s, Peter F. Drucker's seminal publication *Managing in Turbulent Times*[6] expounded the need to address a future that would be shaped by innovation and constant change in a market that has now become global. It is the world of the *New Economy* made of electronic commerce and information technology, which is often transformed into the *Net Economy*, i.e., the economy of communication systems, in particular the exchanges performed through global networks such as the *Internet*. It is, therefore, an economy in which enterprise creation is founded on knowledge rather than on the intensity of financial investments. It is precisely this aspect that facilitates the birth and proliferation of small-sized businesses which then find themselves having to operate in a market characterized by continuous and constant innovation, in other words a market forever undergoing change and therefore highly uncertain.

It is no coincidence that in 1977 John Kenneth Galbraith wrote *The Age of Uncertainty*,[7] while in 1989 Charles Handy defined our era as *The Age of Unreason*.[8] Change, in Handy's eyes, is no longer predictable and gradual: it has become intense and accelerated, discontinuous – and therefore complex and awkward to manage, although far more stimulating. In a context of this kind, the rediscovery of the values that are fundamental in the behavior of the enterprise constitutes the most reliable reference element for entrepreneurial action. Apart from anything else, the entrepreneur is called upon almost daily to perform the role of innovator which, in the teachings of Joseph Alois Schumpeter, is assigned to the entrepreneurial figure at the very moment of enterprise creation.[9]

The situation outlined above is rendered even more apposite to the present-day scenario by the fact that the majority of workers have by now taken on the characteristics of the *knowledge worker*, that is to say, of knowledge professionals who operate on information technology machines – computers – to manage products which are by and large intangible. Defining such products as intangible implies, essentially, that they are composed of information. Moreover, it is more and more the case that these products interlock with the services that are offered to customers along with the products themselves, in an exchange that tends to endure over time. This, however, leaves the enterprise open to constant critical evaluation by the customer, who will not fail to also evaluate the attendant behavior displayed by the enterprise. This means that in order to avoid being penalized, corporate behavior will necessarily have to be played out in a context of positive ethical values.

The "ethical" reference is linked to the semantic meaning of "value," which conveys the proper mode of behavior an enterprise should adopt when going about its business. This implies that the attitudes taken by the enterprise – often conflicting with the expectations displayed by the demand – should be based on appropriate ethical considerations through greater involvement of customers in the entrepreneurial process. Basically, this implies thinking of an enlarged but unitary enterprise network system (macroenterprise) that includes both the upstream elements – the web of suppliers and sub-suppliers – and those downstream – the customers – all working together in an operative context capable of unifying the values they are pursuing.[10]

Certainly, embarking on this path is not without its difficulties, but these are the goals that should be advocated even though the many negative examples arising on the markets would appear to nullify all efforts in this direction. In a book bearing an emblematic title, *Il conflitto epidemico [The Epidemic Conflict]*, Guido Rossi carried out an in-depth analysis of the problem of what values should be pursued, with a very general examination of the conflict of interests that has by now become epidemic on account of widespread acquisitiveness and greed for money. As pointed out by Rossi, this conflict can trigger chaos in the basic structures of the economic exchange markets and on the financial markets as well. In general, the conflict always appears as "a marked imbalance where the scales are tipped in favor of one of the actors."[11] Rossi states "What is needed is not so much an abstract recovery of classical individual morals, restoring to the moral sphere the impact it is felt to have lost, but rather an attempt to engage in a far more concrete operation, namely the introduction [...] of the aspect anthropologists call 'the culture of shame.' "[12] He then adds "In order for there to be some sort of change, in order for the tangle of conflicts of interest that threatens to suffocate our economies to loosen its grip, something has to happen on a different plane: and perhaps that really is the plane of individual and collective ethics."[13]

The hope expressed by Guido Rossi can most certainly be shared, and in fact, as far as entrepreneurial values are concerned, a gradual change in collective and thus also individual consciousness can already be perceived. As noted earlier, despite a degree of resistance and indecision there is a growing tendency to speak of the "social responsibility of the enterprise," where responsibility effectively assumes the characteristics of a mark of quality. This goes hand in hand with rising awareness of the need for a close connection between ethics and economics, which has more than once been proposed as useful and necessary by major thinkers. For example, the problem has been repeatedly addressed by the Indian economist and philosopher, Amartya Sen, who was awarded the Nobel Prize and whose career includes professorships at the London School of Economics, Oxford and Harvard, as well as the position of Master of Trinity College, Cambridge, UK. In support of his arguments Sen goes so far as to formulate stable and well-defined ethical criteria that should from the very outset characterize economic and entrepreneurial decisions.[14]

Overall, then, attitudes are moving toward rediscovery of values as resources that belong both to the individual and also to the enterprise subject. It is a perspective that should be stimulated in large-sized enterprises but also, and above all, in small enterprises. For it is specifically the latter that can be characterized as a preliminary act of intelligence whose absolute protagonist is the entrepreneur:

it is the entrepreneur who proposes the crucial innovation which will act as the springboard for success, to be consolidated through proper interpretation of ethical values.

By combining this conception with the definition of the enterprise as a production network, it becomes possible to forge a connection between the single individual and the second-level individuality constituted by the enterprise itself. In this way, individual values are transferred to the entrepreneurial dimension. Accordingly, the enterprise should be regarded as a social subject, with the entrepreneur as its interpreter. The values that come to the fore in the individual's behavior are thus reproposed in the enterprise, in that the enterprise must bear active testimony to an ethical conception that is sanctioned by the rules of civil co-existence and which is also regulated, at least in part, by law. Moreover, the individual way of thinking is proposed again in the economic framework, which must be preserved in ethical terms as well, so that the relations through which entrepreneurial life will develop can more easily be regulated. The world of ethical values and the realm of economics are thus inextricably intermeshed, shaping an innovative vision that can usher in new systems of social responsibility.

The foundations on which this reform is based rest on a reappraisal of both individual and corporate culture. Ideally, this should be a culture expressed through the concept of quality as an element characterizing management behavior and processes. This entails that it must be a culture constantly underpinned by a search for the best in terms of effectiveness and efficiency, and therefore committed to reaffirmation of the fundamental values of the individual and the enterprise.

In this framework, the enterprise must necessarily interpret its world of moral values as a resource to be made "visible," in contrast to Adam Smith's "invisible" hand. Only in this perspective will it be possible to achieve a significant mode of behavior such as will be able to restore to entrepreneurial action a climate of social effectiveness as well as economic efficiency.

After Adam Smith, the emergence of Taylorism–Fordism almost completely effaced the memory of the "sympathy" underscored by the Scottish economist. The enterprise became a closed world, divorced from its surroundings, in a context which, although defined as the "scientific organization of labor," ultimately had the effect of generating a more and more brutalized and uncultured labor force. The enterprise thus became transformed, metaphorically speaking, into a "castle," to use a term dear to Federico Butera[15] – a castle where a sort of "baroque court" came into being, rich in intrigue and personal egoism, a court that even today has not been completely swept away. This picture, which was grim indeed, continued to set the scene until the late decades of the 20th century and began to crumble only with the advent of the great revolution ushered in by the introduction of advanced technology into the world of work.

Today we are still in the midst of this revolution, of which we are conscious – but frequently also unwitting – protagonists. It is a revolution that has imposed a new unit of socioeconomic investigation in factories, and more generally in the enterprise, composed as it is of that "strange pair" made up of the human individual and the computer.

The legislation that regulates corporate behavior has yet to move toward this new situation. It is often excessively complex and confused, and it therefore cannot always assure an appropriate framework for management. And in any case it always rests on a punitive approach, never on a positive approach. It would perhaps be helpful, in this regard, to take a fresh look at the Enlightenment teachings of Denis Diderot (1748–1832), in particular as expressed in his notes to the translation of Shaftesbury's celebrated work *An Inquiry Concerning Virtue and Merit*, or at the writings of the Italian thinker Melchiorre Gioia (1767–1829), whose treatise *Concerning Merit and Rewards* opened with the following statement: "Many hundreds of works deal with crimes and penalties; only a few focus on merits and rewards [because] men in general are more inclined to punish than to reward: the punishment is intensely desired on account of eagerness for security, while the reward is opposed by interest and vanity [. . .]."[16]

These are words on which it is fitting to reflect once more today, if we seek to rediscover the complete individual, endowed with moral qualities as well – if, in other words, our aim is to conceive of a society of persons and organizations who, taken together, are capable of interpreting an entrepreneurial existence worthy of being lived, worthy in its zest for adventure and the search for profit, but worthy also in its quest for a renewed system of social responsibility that will prove capable of establishing a new grand alliance between moral values and economics.

This alliance must be stimulated and implemented by a renewed enterprise spirit, springing from a cultural background that can inspire not only the individual operator but also the entrepreneurial unit as an expression of second-level autonomous individuality.

Notes

1. Stefano Salis, *Nuova cittadinanza d'impresa* [*New Enterprise Citizenship*], in *Il Sole 24 Ore*, 8 February 2003.
2. Adam Smith, *The Wealth of Nations*, 1776, http://www.adamsmith.org/smith/won-b1-c4.htm
3. *Dizionario di Economia Politica* [*Dictionary of Political Economy*], ed. Claudio Napoleoni, Edizioni di Comunità, Milan, 1956, p. 1675.
4. *Grande Dizionario della lingua Italiana* [*Complete Dictionary of the Italian Language*], ed. Salvatore Battaglia, Utet, Torino, 2002, vol. XXI, p. 644.
5. Fritjof Capra, *The Web of Life: A New Scientific Understanding of Living Systems*, Harper Collins, London, 1996.
6. Peter F. Drucker, *Managing in Turbulent Times*, Harper Collins, New York, 1980.
7. John K. Galbraith, *The Age of Uncertainty*, Houghton Mifflin, Boston, MA, 1977.
8. Charles Handy, *The Age of Unreason*, Arrow Business Books Ltd., London, 1989.
9. Joseph A. Schumpeter, *History of Economic Analysis*, Allen & Uwin, London, 1954.
10. Pietro Romano and Paolo Danese, *Supply Chain Management. La gestione dei processi di fornitura e di distribuzione [Supply Chain Management: Management of Supply and Distribution Processes]*, McGraw-Hill, Milan, 2006.
11. Guido Rossi, *Il conflitto epidemico* [*The Epidemic Conflict*], Adelphi, Milan, 2003, p. 21.
12. Ibidem, p. 138.
13. Ibidem, p. 25.
14. Amartya Sen, *Development as Freedom*, Alfred Knopf, New York, 1999.

15. Federico Butera, *Il castello e la rete. Impresa, organizzazioni e professioni nell'Europa degli anni '90* [*The Castle and the Network: Enterprise, Organizations and Professions in 1990s Europe*] cit., Angeli, Milan, 1990.
16. Melchiorre Gioia, *Del merito e delle ricompense Trattato storico filosofico* [*Concerning Merit and Rewards: Historical-Philosophical Treatise, 1818–1819*], Third Edition, Lugano, Printed by Giuseppe Ruggia e Compagni, 1832, Preface.

Chapter 9
The Enterprise Spirit and Culture in the Network Enterprise

Corporate culture, which in the age of Taylorism–Fordism referred exclusively to corporate history, is today tending to become increasingly strategic and widely distributed, since it is expressed, as we have seen, by all the individuals who make up the social context of the network enterprise. Professional training is likewise evolving from mere transfer of purely technical knowledge to genuine forms of education characterized by cultural awareness and general knowledge.

The new network enterprise, typically embodying forms of high-tech knowledge, is constantly in search of a cultured individual capable of governing so-called smart machines and interpreting corporate *governance* and the role of self-entrepreneur in the best possible way. Culture thus spreads from the entrepreneur to the management, to the administrators, even to the owners, and more generally to all the stakeholders of the enterprise.

The network enterprise thus takes shape through a social context that is the expression of a rather special community of individuals: namely, individuals who, through their behavior, can express a culture which then also takes on a specific ethical physiognomy. In this perspective one may also point to a rediscovery of the importance of the overall concept of quality, in the sense that within the network enterprise the attention to individual quality becomes integrated into processes that are typical of global quality (*Total Quality*).

Along with these developments, it is interesting to note the rise of a management culture that is increasingly moving toward intelligent use and optimal conservation of products. Thus maintenance processes are taking on a new role, identifiable as the transfer of a specific conserving culture that can also make use of advertising and mass communication tools to achieve particular objectives.

What this implies is that the enterprise must devise a new process of cultural education and training addressed to its customers. Naturally, such a process also involves the ethical plane, thereby opening up new frontiers and hitherto uncharted territory for organization theory. This represents one of the major challenges of today's new revolution as it looks toward the future, where the hidden connections between economic thought and organization theory will once again come into their own by synergically enhancing the two fields of investigation.

It is clear that the enterprise, a tool that makes it possible to undertake an ever greater range of new activities, will increasingly have to take on the role of an act

G. Dioguardi, *Network Enterprises*, Innovation, Technology, and Knowledge Management 104, DOI 10.1007/978-1-4419-1333-3_9, © Gianfranco Dioguardi, 2010

of preliminary intelligence. It will thus be the cornerstone of an innovation that will constitute the basis for success and hence for the production of new wealth. This was aptly stated by Carlo Cattaneo when, in 1860, he wrote that "There can be no labour, there can be no capital but that it starts with an act of intelligence";[1] the same concept was reasserted in his statement: "a person's willpower is a commencement of wealth just as much as is his intelligence."[2]

In seeking to inquire into the nature of the "enterprise spirit," we must realize that the key lies in the intelligence that constantly guides the action of the enterprise. However, intelligence should always be accompanied by additional components which, although not always easy to define, nevertheless contribute to shaping the entrepreneur and consequently also the spirit that is at all times the characteristic of the entrepreneur.

The enterprise spirit follows the development of entrepreneurial life, which can be represented metaphorically by the seasons of the year.

The hallmark of spring, when the enterprise spirit is still in its infancy, is an entrepreneurial innovation that quite frequently arises purely by chance and then enters into the world of necessity precisely through the enterprise. Summer, or life in one's prime, is embodied by the organizational dimension which defines the Enlightenment-inspired rationality of the organizational rules. The fall, the age of maturity, sees the enterprise reaching the phase of accomplished development, but it is also the phase that betrays the first faint signs of the encroaching baroque court, with modes of behavior that are at times egoistic and not oriented toward common goals. The winter of old age brings crises that are not infrequently irreversible unless the wellsprings of new opportunities are replenished, in such a manner as to reinvigorate the entire life cycle and restore effectiveness.

In any case, the enterprise spirit is initially characterized by creativity steeped in freedom, which is manifested in the first place through the process of innovative inventiveness. The creation of a new enterprise is always an act of love, a dream, a risk, often a gamble. Such feelings are eloquently expressed by Marco Vitale in a book that addresses precisely this topic, entitled *Sviluppo e spirito d'impresa [Development and the Enterprise Spirit]*, where among other things the author offers some insightful reflections on a fine passage by Alexis de Tocqueville, taken from his work *Democracy in America* (1835-340): " 'The real cause why Americans can sail faster across the Atlantic than anyone else and keep such low prices is explained as follows: [...] one would seek in vain the causes of this superiority [of the Americans] in material advantages; it is due to purely intellectual and moral qualities [...]' The idea of newness is [...] in its spirit intimately linked with the idea of the best."[3] Vitale also points out that the enterprise spirit should be regarded as consisting in the intellectual and moral qualities individuals are endowed with, to which should be added a strong drive toward innovation that is always accompanied by the sense of excitement that comes with striking out into the unknown – in other words, the adventure spirit.

But what Vitale particularly seeks to investigate is the spirit of the type of enterprise that is particularly dear to his heart: an enterprise that is always at the service of mankind and can compete with unruffled serenity and freedom. Such an enterprise features three accumulation processes: "accumulation of capital, organizational knowledge, and technological knowledge."[4] Its aim should be that of contributing to development and – with Schumpeter in mind – always acting as the fountainhead of innovations in the economic system. And here innovation is interpreted not only as a science or a technology, but above all as a "value." Thus there is also a return to the concept that the healthy enterprise is a working organization, a society of men and commodities gathered together around a legitimate project. The aspects that should ideally come to the fore in this organization concern not only a range of professional skills and competencies but also a culture capable of ennobling the participants by unifying their actions through a unitary enterprise spirit. And so, "the enterprise is not a set of rules, equations, techniques, amounts of capital, but a society of men united and integrated around a common project that is in continuous evolution,"[5] "which by its very nature is based on development of man's intellectual and moral capacities,"[6] and again "enterprises are not only the place where production is organized and income is accrued, but also the place where productivity is generated and the values functional to the development of productivity and of human labor are shaped, a place where culture is created."[7] This is indeed a vision of the enterprise strongly shaped by the humanist spirit, and since it must necessarily be projected onto the scenario in which it operates, if it is to remain true to this spirit, then the scenario must be that of an open and just society, rational and not feudal, not blighted by a criminal mentality.

From an organizational point of view, the enterprise is fundamentally the place where governance over the various different modes of corporate behavior is manifested. According to Lucian Gallino, "governing an enterprise means determining its production strategies; the complex financial and organizational architecture; the allocation of profits between capital and labor; the size and nature of investments. It also means deciding what mergers and acquisitions are suitable in order to make the enterprise grow, or whether it might not be the right moment for divestiture of sectors held to be unproductive; what lines of activity to take up or abandon for purposes of diversification or integration. Last but not least, it entails appointing or dismissing top executives. In this sense, corporate governance has always existed, whatever the label it happened to be known by."[8]

But on the other hand, the enterprise spirit needs to be rediscovered in the very thrill of creating, doing, embarking on action, achieving. The truly professional approach, understood as an indispensable value for the civil development of society, must always be accompanied by cultural awareness, through which man can recognize himself as master of his own destiny. This spirit, which is abundantly present in small enterprises, is however often absent in large corporations. In this regard Marco Vitale draws a contrast with the conception that is still present in many large enterprises, where the dominant models are based on power and corruption together with political protection and embezzlement. He proposes instead that one should take a fresh look at small enterprises, generally family run, where ownership and management tend to coincide, that is to say family businesses where the family itself represents the capitalist and is therefore the intelligent owner.

This also underscores the great moral strength that must characterize the entrepreneur so that it can then be transferred into the enterprise. In this vision, the enterprise appears as a group of individuals oriented toward the achievement of common objectives, often capable of building up an internal environment from which an ethical context must necessarily be made to spring forth. It will be a morally motivating environment, in order to facilitate the coexistence of people who have to work together and to enable them to be more satisfactorily guided toward ever greater operative effectiveness and efficiency. These individuals will thus be prompted to interpret the role of self-entrepreneurs, pursuing and bringing into play the very same moral endowments that characterize the entrepreneurial leader. Furthermore, such an environment will be conducive to the formation of an atmosphere of constant cognitive learning, through which the cultural interests of individuals will contribute to shaping a satisfying corporate culture that will also be motivating in the ethical sense. Culture and ethics, then, are both destined to be transferred to the external environment, not only through the individual workers but also by virtue of the overall behavior of an enterprise that can now be considered as socially highly developed.

Moreover, in the eyes of its surrounding local area the entrepreneurial organization itself appears as a subject capable of expressing a typical second-level individuality which presents itself as the true embodiment of the enterprise spirit, with its culture, tradition, history, and ethics springing from manifestations of its spirit.

In short, social responsibility in the sense of the value system to be built up within the enterprise is becoming an increasingly widespread phenomenon today and is thus attracting considerable interest and awareness. But it has venerable and long-established origins. Mention was made earlier of Adam Smith, the founder of modern economics, who rose to instant fame in 1776 when the publisher William Strahan brought out the first edition of A*n Enquiry into the Nature and Cause of Wealth of Nations.* But Adam Smith was also the author of an earlier publication, dating from 1752, namely another fundamental book bearing a significant title: *The Theory of Moral Sentiments, or An Essay towards an Analysis of the Principles by which Men Naturally Judge Concerning the Conduct and Character, First of their Neighbours, and Afterwards of Themselves*. In effect, the economic analysis so insightfully conducted in *The Wealth of Nations* had been preceded by studies on the values of moral behavior, so that the world of business – dominated by the egoism of human action – could likewise be built on the presupposition of an innate "sympathy" inherent in human beings. And this sentiment of sympathy was introduced as follows: "How selfish soever man may be supposed, there are evidently some principles in his nature, which interest him in the fortune of others, and render their happiness necessary to him, though he derives nothing from it except the pleasure of seeing it. Sympathy [...] may now [...] without much impropriety, be made use of to denote our fellow-feeling with any passion whatever."[9] Historically, this marked the beginning of an ever more tightly meshed connection between the individual's disposition toward ethical values and the phenomenon of economic behavior. Once more, this enabled Smith to theorize that in a world dominated by the egoism of human action, man's actions – guided in their statistical aggregation by a sort of "invisible hand," would often lead to results that were useful for the collectivity. The same concepts have recently been put forward again by Alfred Chandler in his book entitled – rephrasing an allusion to Smith – *The Visible Hand*, referring to the explicit actions performed by the managers of an enterprise. Such actions are capable of exerting visible influence on the external environmental scenario, in particular on the market, precisely through the behavior the managers have identified and set in place in the course of their entrepreneurial action.[10]

But, as noted earlier, Smith himself, in his *Theory of Moral Sentiments*, had previously pointed to the innate existence of a sentiment of "sympathy" in the individual, arguing that sympathy was capable of influencing, and indeed attenuating, human egoism. This sentiment present in the single individual must be enabled to transfer itself, through a natural process, to corporate behavior as well. Such, in fact, is the teaching of Alfred Chandler in our own time, where he refers to the explicit action performed by enterprise managers, who can visibly influence the external environmental scenario, in particular the market, precisely through the type of corporate behavior they have built up and set in motion: "The theme propounded here is that modern business enterprise took the place of market mechanisms in coordinating the activities of the economy and allocating its resources. In many sectors of the economy the visible hand of management replaced what Adam Smith referred to as the invisible hand of market forces. The market remained the generator of demand for goods and services, but modern business enterprise took over the functions of coordinating flows of goods through existing processes of production and distribution, and of allocating funds and personnel for future production and distribution."[11]

Thus the enterprise must necessarily act on the basis of an ethical vision that allows it to attain a socially significant mode of behavior, but it must be a vision which is capable of regenerating economic efficiency at the same time, through entrepreneurial action. The picture thus delineated is one that Federico Butera defines as the "socially capable enterprise," that is in a position to bring about "products, services, processes, but above all real people who have grown up and become socially fulfilled in the enterprise and with the enterprise: managers, professionals, technical personnel, craftsmen, simple workers, and also customers and suppliers [...]."[12] Furthermore, "The socially capable enterprise" generates wealth not only for itself, as it also draws suppliers and customers, local communities, and global systems into this process. "The socially capable enterprise" fertilizes communities, local economic systems, public administrations, in addition to other enterprises as well, and accumulates over time a substantial "social capital." It represents a factor of competitive advantage and almost a predictor of success in the economic sphere as well.[13] This is in sharp contrast with the concept of the "irresponsible enterprise" theorized by Luciano Gallino: "An enterprise is defined as irresponsible when it believes that apart from the elementary obligations laid down by law it is not required to answer to any public and private authority, or to public opinion, with regard to the consequences of its activities as they affect the economic, social and environmental field. Such consequences include: industrial and financial strategies, the labor conditions imposed on its workers in its own country and also abroad; employment policies; the relation of its products and production processes to the environment; the utilization of funds that have been entrusted to the enterprise by savers in the form of shareholdings or bonds; the drawing up of statements of account; the localization or delocalization of production activities; its tax behaviour."[14]

What is made clear by these observations is that the enterprise should avoid being overcome by the slavish pursuit of mere economic profit as the one and only aim of corporate management. That is to say, the enterprise spirit which Cattaneo portrayed as an act of intelligence should not be downgraded by reductive and purely egoistic conceptions. Rather, the enterprise spirit should be intent on rediscovering the individual, complete with the human being's moral dimensions. The spirit defined in these terms should be capable of building a society of individuals who can collectively interpret an entrepreneurial existence that is worthy of being experienced – worthy in its zest for adventure and profit, but also in its awareness of motivations of an ethical nature.

This signals the end of the era in which the division of labor was designed exclusively to boost productive efficiency and consequently to increase the quantity produced. That was the scenario of a bygone era, which inevitably entailed pronounced specialization in operative professional skills, regardless of any general cultural context. Culture in this earlier system was held to be virtually useless for the purposes of specialist and professional training: instead, training was to be developed in the framework of extreme fragmentation and splitting up of the various working tasks.

In contrast, the picture now being forged is that of a new enterprise spirit, expressed by a set of persons who have the ability to counter the egoism of the baroque court. The motivation that must represent their driving force is the pursuit of common objectives in a socially significant context, which is in effect made up of the whole range of individuals who work within the enterprise. This implies none other than the rediscovery of the personality of the human being, thus also allowing the rationality that is part and parcel of the human personality to come to the fore once again, and directly reviving the memory of the frame of mind and background of knowledge that has come down to us from the Enlightenment.

This new enterprise spirit is capable of generating a socially motivating internal environment that draws to some extent on principles of behavioral ethics. Such a framework provides the inspiration that underpins coexistence among persons who have to work together, in constant interaction with high-tech machines, and guided by ever greater operative effectiveness and efficiency in the pursuit of common objectives. Accordingly, the prevailing atmosphere has to be oriented toward continuous cognitive learning, the end result of which is the rise of a new corporate culture. With it comes a new entrepreneurial ethics, transforming the core of the enterprise into a veritable new "social laboratory" of cognitive experimentation.

As we have seen, culture and ethics are subsequently destined to be transferred to the broader surrounding area that has entered into the sphere of influence of the enterprise. This comes about first and foremost because the enterprise has succeeded in amplifying its organizational structure by developing outsourcing relations with the enterprises forming part of its web of suppliers and sub-suppliers (macroenterprise). Moreover, since the individual coworkers now enjoy a greater availability of leisure time, they can live more intensely in the external environment, influencing but also undergoing the influence of their new contacts. It might even be argued that they act as latter-day "urban missionaries," spreading the cultural and ethical teachings acquired in the corporate *milieu*, from which they constantly draw the inspiration for regeneration in a relation of reciprocity.

In parallel, the concept of *stakeholders* is likewise being extended. Taken in the sense of all those whose interests bring them into the sphere of influence of an enterprise, the stakeholders now interact with an enterprise that is capable of transmitting important and widely diversified values, starting from the simple and exclusive pursuit of profit, which must in any case remain the fundamental element on which to build up the boundaries of the new frontier. Crucially, though, it is a frontier that can move beyond the egoisms of the baroque court and is ready to rediscover the dimension of sympathy evoked by Adam Smith. This new approach is attributable to a mode of ethical behavior that cannot but lead to noticeable and significant advantages not only for the individual operating subject but also for the social community composed of the "laboratory-enterprise" and for the surrounding area where the enterprise is active and builds up interactive relations.

Notes

1. Carlo Cattaneo, *Scritti economici* [*Economic Writings*], vol. III, Le Monnier, Florence, 1956, p. 344.
2. Ibidem, p. 361.
3. Marco Vitale, *Sviluppo e spirito d'impresa* [*Development and the Enterprise Spirit*], Il Veltro, Rome, 2001, p. 329.
4. Ibidem, p. 117.
5. Ibidem, p. 130.
6. Ibidem, p. 132.
7. Ibidem, p. 130.
8. Luciano Gallino; *L'impresa irresponsabile* [*The Irresponsible Enterprise*; cf. also "Managerial Irresponsibility and the New Business Model," in *Cultura e Impresa* [*Culture and Enterprise*], online journal, http://www.cultureimpresa.it/04-2006/english/primo01-en.html], Einaudi, Turin, 2005, p. 116.
9. Adam Smith, *The Theory of Moral Sentiments*, Part 1, Chapter 1, http://www.adamsmith.org/smith/tms/tms-p1-s1-c1.htm
10. Alfred D. Chandler, *The Visible Hand: The Managerial Revolution in American Business*, The Belknap Press, Harvard, Cambridge, MA, 1977, p. 1.
11. Ibidem.
12. Federico Butera, *Il campanile e la rete. L'electronic business e le piccole e medie imprese in Italia* [The Belltower. . .], Il Sole 24 Ore, Milan, 2001, p. 67.
13. Ibidem.
14. Luciano Gallino, *L'impresa irresponsabile* [*The Irresponsible Enterprise*], cit., vii.

Chapter 10
Enterprise Networks

Enterprise networks give rise to new economic entities of investigation that can be considered as unitary third-level forms, distinct from persons taken as expressions of first-level entities and from enterprises taken as second-level individualities. In these third-level aggregations the concept of network is once again helpful in building up a shared atmosphere among the enterprises that belong to the aggregation. It is an atmosphere always characterized by operative aspects in the sense that it induces individual forms of behavior that resemble one another and are thus able to build up a common culture.

In the past, enterprise networks were typically a phenomenon involving a local area. Thus they tended to be expressed as districts, with their various forms of industrial agglomerations. The industrial districts subsequently developed into technology districts and science parks. More recently, networks have begun to take on forms that pursue functional ends, as in the case of somewhat casual and unintentional aggregations in the "macroenterprise" form, i.e., in the system of enterprises that make up the web of interconnected suppliers and sub-suppliers of a larger enterprise, whereby the latter act as a natural focal point for the surrounding enterprises.

But even in the macroenterprise one notes a shared atmosphere that grows up around the various enterprises in the supply chain and molds their attitudes. Fundamentally, in this case the atmosphere is shaped by the intentions flowing from the reference enterprise and the contractual relations the latter sets up. This is particularly true for contracts that regulate transactions with the individual enterprises. The services and tasks specified in the transactions are bound by quality and delivery time constraints which have the consequence of unifying the entrepreneurial behavior of all the enterprises involved. If such constraints are not respected, there may be adverse effects that have direct repercussions on the final product and impair the quality pursued by the reference enterprise. It follows that the reference enterprise must be fully involved in building up a virtuous atmosphere, constantly aiming to work toward a quality, and more generally a culture, which will harmonize its objectives with those of the individual enterprises operating within the macroenterprise. Only thus can the desired "network effect" be truly generated.

The macroenterprise network must also encourage research and training, enhancing the stimuli that spring from the reference enterprise which often acts as a guide

G. Dioguardi, *Network Enterprises*, Innovation, Technology, and Knowledge Management 104, DOI 10.1007/978-1-4419-1333-3_10, © Gianfranco Dioguardi, 2010

in orienting the various actions. At the same time, however, the individual enterprises belonging to the macroenterprise need to be allowed sufficient independence to solve problems that may arise from time to time, as this flexibility offers scope for the development of widespread innovation.

Looking beyond the perspective of the macroenterprise, it can also be observed that in some cases enterprises – in particular small- and medium-sized enterprises – may, with the aim of enhancing their potential, join together in various forms of consortiums (joint ventures, temporary associations, true consortiums, and so forth). This leads to the rise of further external enterprise networks, whose functional logic is in this circumstance dictated by the individual entrepreneurial decisions aimed at developing well-defined associative strategies that are designed for the specific purpose of setting up the new aggregation. Each of these associations has to shape a special atmosphere of its own, since this is indispensable in order for the aggregation to be able to coordinate its actions and thus operate successfully. That is to say, even though the actions are expressed by the individual corporate entities, they are coordinated in such a manner as to become the unitary expression of the new third-level entrepreneurial unit that is represented precisely by this form of association.

Moreover, in the present general context further new forms of entrepreneurial aggregations are emerging. They consist of enterprise networks that stand half-way between those with functional characteristics as already described and the traditional local networks expressed by the industrial districts. It is worth noting, however, that the rise of this kind of network, in a sort of virtual context, is almost always prompted by problems linked to the small size of the enterprise.

This is in fact an important observation, because the most representative size in all economic contexts is that of the small enterprise. And this is particularly true in Italy, where it is precisely the small enterprises that have achieved success, exporting *Made in Italy* brands on international markets. According to Giangiacomo Nardozzi, "The extensive spread of industrial techniques among small enterprises allowed these smaller units to gain certain advantages as compared to the large plants of major companies, such as flexibility in production volumes and in utilization of labor. These advantages were of enormous benefit as regards the ability to adapt under conditions of increasingly volatile demand and growing labor market rigidity. Thus there was a thrust towards bottom-up locally rooted industrial development, and local production was able to make the most of a devaluation that had been calibrated on the production conditions of the large companies. The small enterprise became a success story."[1]

This situation should be considered as a phenomenon to be encouraged rather than opposed. It is a fallacy to assume that the main objective should be the race to larger corporative entities at all costs, as this entails the risk of forging huge entrepreneurial dinosaurs which, however, rest on feet and structures made of fragile clay.

Nevertheless, precisely the small-sized enterprises are those in greatest need of being fuelled by continuous and constant innovation, and frequently they are unable to achieve this goal on the own, with the result that they face the threat of being driven out of the market as soon as the initial creative business idea becomes obsolete. But it is here that the network configuration comes into play: innovation can be favored by the web of exchanges that the network is always capable of assuring and facilitating.

However, other potentially negative aspects also have to be addressed. Today's markets are global, and their demand for products is constantly expanding from the quantitative point of view. This has been observed in the well-known case of the American markets, and also in the emerging markets of the Far East, with special reference to the Indian and Chinese markets. Mario Deaglio has offered the following comments on this aspect: "The years after the turn of the 3rd millennium – when the long phase of expansion of the American economy had run its course and 'market globalization' began to wane – showed a pronounced shift in the weighting of production in the various areas of the planet. A particularly striking phenomenon was the extremely rapid growth of the Far East Asian area. [...] almost half [...] of the new production of the planet originates from the dynamic and advanced areas of the Far East, namely from China and the other so-called 'Asian tigers', including India as well as Japan. For the sake of simplicity this area will henceforth be referred to here as 'dynamic Asia', even though the Japanese economy does not particularly stand out for its dynamism at the moment."[2]

The result is that on the one hand the picture of Italian enterprises shows congenital dwarfism which is, however, highly innovative and qualitatively significant, while on the other hand the dimensions of demand are increasing and demand on the new markets is always enormous.

It is therefore imperative to take action in order to provide support for continuous innovation that will go hand in hand with traditional *Made in Italy* quality. Only by appropriate backing of this kind will it be possible to sustain interest in Italian products on world markets, whatever the dimension of the Italian production. In this respect, it is interesting to call to mind the Schumpeterian conception of the enterprise, namely, as an innovative undertaking to be maintained over time through innovation and to be constantly enhanced in its quantitative and qualitative dimensions. Such a conception is even more appropriate today, when the scenarios of turbulence and frenzy foretell rapid changes affecting all the rules of the economic game, not to mention the incessant oscillations of the exchange markets. But far from presaging gloom and doom, such changes call for a positive response that makes full use of a network, for not only is the flair for innovation an inbuilt characteristic of many enterprises belonging to a network, but the very presence of a network facilitates the circulation of innovations.

The themes linked to the concept of innovation, and the problems that generally arise in the case of small-sized enterprises, can in fact be more easily addressed by reference to the organizational concept of an external entrepreneurial network. This kind of network is in a position to assure specific operative flexibility while not eschewing the search for greater and significant size. But it is important to realize that what is involved here is a rather innovative conception of network enterprise that can express itself both internally and on the outside through a network of enterprises linked to one another by clearly targeted agreements. Implementation of these agreements is carried out in reference to the markets the enterprises wish to penetrate, operating in forms that guarantee new localisms on the emerging areas in question, and thereby also introducing a feature which is innovative from the commercial and not merely productive perspective.

In short, the old Marshallian industrial district – another typical *Made in Italy* element – today must be restated in terms of a radically innovative concept of novelty; thanks to the IT revolution and the advent of high-tech. In today's world, the district steps beyond local or territorial boundaries to become the object of immaterial infrastructuring – generally linked to the city context – thus giving rise to what Carlo Trigilia, in a book on *Sviluppo Locale, un progetto per l'Italia,* calls "the high-tech districts," whose hallmark is in fact the type of innovation expressed by advanced technology. Their existence is justified – as Trigilia points out – by the fact that "[...] innovative activities do not generally arise in an isolated manner [. . .] What they require most of all are effective formal and informal relations between the university world and the world of research, enterprise systems, specialized financial services (*venture capital*)."[3] In addition, as Trigilia states elsewhere, "Over the past few years attention has often focused on a different type of local production system, namely a concentration of small enterprises in sectors considered to embody high-tech activity. Among these, a particularly important role is played by software production, biotechnology, and media-related productions (movies, television, etc.). In fact, these concentrations are often referred to a 'high tech districts'. And the phenomenon has given rise to a curious paradox. While on the one hand there is widespread concern as to whether the districts will really be able to survive delocalization processes, on the other hand certain sectors of these new technologies, facilitated by new communications technology, actually seem to be developing into forms of local production systems composed of small enterprises."[4]

Thus what we have in this case is once again a network of enterprises which can aptly be termed a *web,* whose reference node – or spider at the center of its web – must be conceived in innovative terms as a veritable pole of excellence, acting as the expression of a multiplicity of interests. The most immediate referent that comes to mind is that of universities and research institutions, but the network of interconnected enterprises constituting a pole of excellence can also be the expression of local institutions, management consultancies, and the higher advisory echelons of entrepreneurial organizations, and perhaps enterprise consortiums as well. In effect, one has to imagine setting up specific poles of excellence for each specific phenomenon of entrepreneurial localism, even that of a virtual type, so as to restore to each district and high-tech network its own technological identity. It is these poles that must act as the wellspring of organizational governance, but also of research, innovation, advanced training. They must be capable of expressing an excellence that will elicit a flow of innovation, just as used to be the case with the former concept of the district, so that enterprises can rise to the challenge of the great new markets in the era of globalization.

Another important aspect that should by no means be overlooked concerns provision of assistance for the internationalization of enterprises. This aspect should be vigorously pursued through the network mold that has been forged during the quest for new operative synergies. Help should be made available in the form of an analysis of interesting opportunities on the new emerging markets, encouraging enterprises to enter such markets. In this framework, it is possible to work toward the aim of reaching sufficiently significant dimensions for entry by setting up specific agreements among enterprises, even among enterprises that do not share a common local geographic area if they appear interesting from the point of view of product integration or complementariness. This effectively leads to the formation of a sort of "virtual entrepreneurial district" which allows the diverse enterprises to present themselves as a united entity on the new markets.

This perspective can be further advanced by encouraging enterprises to form associations, so that they can pursue the goal of achieving a more solid presence on the markets. Therefore, support should be made available for the establishment of networks that represent the expression of entrepreneurial groups comparable to *clusters*, or "virtual districts." Such groups would include not only small- and medium-sized local units that share similar market characteristics but also enterprises from other districts with analogous or at least complementary characteristics, in cases where these enterprises are interested in achieving an adequate presence on large markets such as the emerging markets. A prospect of this kind requires the construction of enterprise networks capable of expressing corresponding activity networks appropriate for an overall strategy designed to boost the growth of the local entrepreneurial context.

The measures introduced to enhance internationalization and innovation need to be integrated with an intense advanced professional training process, addressed first and foremost to the entrepreneurs themselves and then to all those working within the enterprises. This could be done by setting up purpose-designed study and research seminars on economic and social themes of interest to the enterprises involved in these forward-looking activities. Ideally, such seminars could also serve the more general purpose of arousing curiosity and prompting the development of ideas and projects that would enhance professional skills and contribute to lifelong education.

Thus the enterprise network, that has so far been represented on the local level by the classical industrial district, must find a means of transforming itself into a new unit of economic investigation, a district that is, so to speak, "virtual." It will include commercial agreements among enterprises that may be geographically distant from one another, yet share similar features in terms of quality of production, and such enterprises will establish links with other usefully complementary enterprises in order to build up a virtuous network that can operate as an integrated corporate system. Its great advantage will lie in its ability to aggregate the aspect of multiplicity – with its valuable connotation of quality – that is an intrinsic feature of small size, so that the resulting unit is ready to engage in trade and exchange in terms that are significant from a quantitative point of view. Therefore, local development must be buttressed by enterprise networks that operate as veritable enterprise systems, and to this end it is vital that they should be able to enjoy constant interaction with a network of poles of excellence.[5]

The new forms of high-tech aggregation should be regarded as an evolution of the traditional technology districts. Indeed, one of the important aspects of enterprise networks is that they can truly be described as "technology parks." This definition is generally applied to sectors in which technologically advanced enterprises operate, and such enterprises are given access to high technology services in order to optimize their management and ensure their survival and most suitable development.

The services provided in the framework of technology parks must, in the same way as the enterprises themselves, cultivate a specific climate of entrepreneurial action that will power both economic and social development. Moreover, given their overarching goal of stimulating innovative economic activities and the creation of new productive enterprises, technology parks can in a sense be considered as "enterprise-generating enterprises." This helps to explain why they are often set up on sites with purpose-designed facilities (*incubators*) which are specifically planned to assist new enterprises during their start-up and launch phases.

Clearly, this approach calls for a new strategic vision, which must be guided by forms of entrepreneurial leadership that are best suited to stimulate innovation and internationalization. The contribution and performance of the various figures working in the enterprises will be vital in helping to realize this vision. By the same token, entrepreneurial training will also play a new role, to be redesigned according to the growing organizational requirements so that the opportunities offered by the new frontiers of the global market can be successfully seized.

These are today's new great challenges, and they need to be addressed jointly with the universities, since these are and have always been the home of research and advanced education. Accordingly, the current trend toward the formation of a locally rooted network of universities should be endorsed, while naturally ensuring that they maintain levels of excellence. The ideological basis of these academic institutions should be grounded on the goal of providing ever better service to the local communities: for example, one interesting proposal is that of setting up academies and colleges in the various industrial zones in order to offer a positive response to the real or potential requirements of the local areas. This academic network should, however, constantly maintain the ambition of cultural and specialist preeminence, never yielding to the temptation of purely self-congratulatory proliferation accomplished at the expense of the quality of teaching.

The specialist poles and the centers of excellence must be energized by precisely this kind of network, so that they become the operative expression of the university institutions and can be oriented once more toward that which was their original vocation, namely advanced teaching and research. These poles must therefore become the proactive connections of a new alliance whose mission is to express innovation, new training, and internationalized entrepreneurialism, underpinned by continuous and constant university research and guided by a scientific leadership that can inspire researchers to become enthusiastic protagonists and to instill this same enthusiasm in their operative teams.

The virtual districts, linked to specific poles of excellence that endow them with a distinct character, are in many respects reminiscent of the particular kind of district that has generally become known as the "science park."

The first science parks came into being in the United States in the 1950s. Initially, they arose through spontaneous processes, springing from the concrete initiative of numerous researchers and local entrepreneurs who were interested in promoting commercial applications of the results of their research. A typical case in point is the spontaneous emergence of the entrepreneurial projects in "Silicon Valley," which owed their progress to the innovative research efforts of Stanford University and the entrepreneurial spirit of men such as William Hewlett, David Packard, and Steve Jobs, the founder of Apple. A rather similar phenomenon underlies the entrepreneurial clustering around Boston's Route 128, aided by the driving force of MIT. Analogous projects started up in Arizona and Texas, in the so-called Corridor between Austin and San Antonio. Still others operate in the North Caroline research Triangle, between the cities of Raleigh, Durham, and Chapel Hill.

All these are situations that tend to provoke a cascade effect, as a result of which the science parks, at least in the United States, are assuming considerably significant dimensions. This type of district is proving to be a useful tool for the development and support of a network of industrial enterprises, thus acting as a sound economic scaffold for a country. This can be seen in Japan, where a similar situation has helped to sustain domestic entrepreneurial endeavors and heighten the overall international competitiveness of the nation. Thus in Japan a state program was launched in 1980 for the creation of Technopolis, but in contrast to the US experiences this was established according to a rigid plan set up by the Ministry of Industry and Trade. In 1983 it was transformed into a law by virtue of which 20 areas were designated as sites for the creation of a science park, to be known as "Technopolis." The first to be built were those of Okayama, Hiroshima, Ube, and Kumamoto, situated in the south of the country. All the parks have been operative since the end of 1991.

The science park model arrived in Europe in the early 1970s, where it gave rise to other and different development models. Their focal point was constituted on the one hand by the University of Cambridge in Britain and on the other hand by Sophia Antipolis in France. However, for a fairly long period of time these experiences remained limited to the two countries in question.

In Cambridge attempt was made to construct a technopolitan atmosphere that would foster and bring to life new initiatives in the surrounding area. There were no plans to build specific facilities, but great effort was devoted to the dissemination of research and the training of executives who would assure more effective and efficient management of the various projects.

Sophia Antipolis, by contrast, is a technology park conceived in order to develop – partly and indeed perhaps principally in terms of specific facilities – a vast inland area in the vicinity of the Côte d'Azur. It is located on what was at that time a non-urban greenfield site not far from Nice, and it represents one of the few European cases that aim to combine productive and research activities with the quest to devise a new urban model that allows its inhabitants to experience new forms of productive, social, and cultural aggregation. Overall, it constitutes the largest science park in Europe, occupying a surface of over 2500 hectares and hosting over 700 enterprises, with intense job creation. Among the main factors of success of Sophia Antipolis, a key element is that of cross-fertilization, i.e., the opportunity for exchange of ideas among individuals and organizations, or between enterprises and research institutions, focusing on advanced education and training as well as the creation of nuclei of excellence for research and development activities.

In the 1980s other European countries, above all in Western Europe, began to follow the science park approach, partly as a means to regenerate decommissioned and semi-derelict industrial areas (Bicocca in Milan, Naples, but also Berlin and Birmingham) and partly also to develop newly industrialized areas (for instance, Valencia and Lisbon).

When enterprise networks take shape within the sphere of a specific local area, they express what is usually defined as the classical "industrial district," according to the definition given by Alfred Marshall in the early 20th century and recently taken up again by the school of Giacomo Becattini. It is of considerable interest to read Becattini's observations in this regard: "I define the industrial district as a social and locally rooted entity characterized by the active joint presence, in a particular geographic area that is naturalistically and historically determined, of a community of persons and a population of industrial enterprises. In the district, unlike in other environments (such as the manufacturing city), the community and the enterprises tend, as it were, to become inextricably interwoven (1989)."[6] Other authors, such as Piero Formica and Aldo Maccari, propose the following description: "The characteristic of the *industrial district* is that of being a set of small and medium-sized enterprises, generally (but not always) specialized in terms of the processing phase the given enterprise is concerned with, which operate in a relationship of competition but also of reciprocal cooperation, and having implicit cultural, professional, ethical and behavioral 'rules of the game.' "[7]

On the subject of the industrial districts it is also worth mentioning the comments put forward by Roberto Grandinetti of the University of Padua, as reported by Franco Locatelli in the newspaper *Il Sole 24 Ore*: "[...] what is striking is the fact that the districts which are proving most resilient are those that have within them a multiplicity of formulas and competitive e models, and above all those that have been able to generate enterprises recognized as world leaders, such as the eyewear and spectacles district of Montebelluna, where there are now four enterprises that have risen to medium-large size and whose dynamism almost risks jeopardizing the delicate balance of the district itself."[8] Locatelli also adds that "There is a widespread belief that the district cannot be an unchanging state of affairs and that the more it succeeds in renewing itself over time, the more it will continue to be a living and life-giving model."[9]

Neither the industrial district nor what in Italy has been denominated the *Area* (A.S.I.) can be fully identified with the concept of the technology park, as the latter has a broader range of interpretation. It includes the industrial district and the Industrial Development Area, but extends the type of area, so that it also covers metropolitan areas, where new tertiary sector industries (*software factories*) may be located in city offices.

This shift in perspective with regard to geographic delimitation has interesting implications for the future development of industrial districts and technology parks. Traditionally, the geographic delimitation of industrial districts, based on local areas, could be viewed as representing a catchment basin capable of also using the services supplied by the infrastructures already in place. But this interpretation can now be reformulated and fine-tuned by virtue of an extended construal of the concept of infrastructure: the infrastructure is something that is indispensable if an area is to become suitable for the localization of highly advanced industrial, or more generally entrepreneurial, activities.

The Marshallian districts typically arose, often unplanned, in areas where transport infrastructures had already been set in place (roads, railways, ports, and airports) and primary services were already available (water mains, sewerage, electricity supply, other alternative sources of energy). Such districts were usually not far from the cities, yet at the same time always distinct from them. By contrast, the Industrial Development Areas were designed according to primary infrastructuring programs, with the specific intent of enabling economically depressed and backward areas to become industrially developed. In both cases, the localization was deliberately chosen in such a way as to distance industrial activities, which were often highly polluting, from the urban agglomerations, while taking care to ensure relatively easy access to nearby cities where the other services necessary for development were freely available (hospitals, schools, markets, stores, etc.).

Today, more widespread economic prosperity and the diffusion of advanced technology render primary and secondary infrastructures a necessary but not sufficient condition for development of the local area. In order to promote entrepreneurial activity it is indispensable also to ensure the presence of the kind of infrastructure that is typical of an advanced tertiary sector, which includes efficient cable and satellite connections as well as all the typical high-tech services. Therefore, this is a type of infrastructure that is no longer linked to a specific physical territory: rather, it belongs to an ideal location that may also include the urban agglomeration, where new factories producing non-polluting technological knowledge (*know-how*) can be set up. Such knowledge is typically for programming activities (*software production*) and forms the premise for a type of work that can also be carried out in the residences of the various operators (*working from home*).

Notes

1. Giangiacomo Nardozzi, *Miracolo e Declino, L'Italia tra concorrenza e protezione* [*Miracle and Decline: Italy Between Competition and Protection*], Laterza, Rome-Bari, 2004, pp. 77–78.
2. Mario Deaglio, *Lo spostamento del baricentro economico del mondo* ["The Shift in the Economic Center of Gravity of the World"], in Mario Deaglio, Giorgio S. Frankel, Pier Giorgio Monateri, Anna Caffarena, *Il sole sorge a Oriente* [*The Sun Rises in the East*], Guerini, Milan, 2005, pp. 5–6.
3. Carlo Trigilia, *Sviluppo locale: Un progetto per l'Italia* [*Local Development: A Project for Italy*], Laterza, Rome-Bari, 205, p. 182.
4. Carlo Trigilia, *Distretti industriali e distretti high tech* ["Industrial Districts and Hi-Tech Districts"], in *Economia senza gabbie, Studi in onore di Giacomo Becattini* [*Economics Without Cages: Studies in Honor of Giacomo Becattini*], ed. Nicolò Bellanca, Marco Dardi and Tiziano Raffaelli, Il Mulino, Bologna, 2004, p. 29.

5. Gianfranco Dioguardi, *I sistemi organizzativi* [Organizational Systems], Bruno Mondadori, Milan, 2005.
6. Cited by Renzo Bianchi in *Giacomo Becattini e la «causa finale», della sua ricerca sui distretti industriali* ["Giacomo Becattini and the 'Final Cause' of His Research on Industrial Districts"], in *Economia senza gabbie...* [*Economics Without Cages*], cit., p. 17.
7. Piero Formica-Aldo Maccari, *Dal distretto industriale al territorio tecnopolitano* ["From the Industrial District to the Technopolitan Area"], in *Zenit*, 3, 1989, p. 16.
8. In Franco Locatelli, *Si vince solo con l'innovazione e le reti* ["A Real Win Comes Only from Innovation and Networks"], in *Il Sole 24 Ore*, 25 April 2005, p. 5.
9. Ibidem.

Subject Index

Name Index

MIX
Papier aus verantwortungsvollen Quellen
Paper from responsible sources
FSC® C105338

If you have any concerns about our products,
you can contact us on
ProductSafety@springernature.com

In case Publisher is established outside the EU,
the EU authorized representative is:
Springer Nature Customer Service Center GmbH
Europaplatz 3, 69115 Heidelberg, Germany

Printed by Libri Plureos GmbH
in Hamburg, Germany